THE

THERMAL MEASUREMENT
OF ENERGY.

THE
THERMAL MEASUREMENT
OF ENERGY.

LECTURES DELIVERED AT
THE PHILOSOPHICAL HALL, LEEDS

BY

E. H. GRIFFITHS, M.A., F.R.S.

FELLOW OF SIDNEY SUSSEX COLLEGE, CAMBRIDGE.

CAMBRIDGE:
AT THE UNIVERSITY PRESS.
1901

CAMBRIDGE
UNIVERSITY PRESS

University Printing House, Cambridge CB2 8BS, United Kingdom

Published in the United States of America by Cambridge University Press, New York

Cambridge University Press is part of the University of Cambridge.

It furthers the University's mission by disseminating knowledge in the pursuit of
education, learning and research at the highest international levels of excellence.

www.cambridge.org
Information on this title: www.cambridge.org/9781107424067

© Cambridge University Press 1901

First published 1901
First paperback edition 2014

A catalogue record for this publication is available from the British Library

ISBN 978-1-107-42406-7 Paperback

PREFACE.

THE following Lectures were delivered in Leeds, during the Spring of this year, at the request of the Technical Instruction Committee of the West Riding County Council.

At the close of the course I received, from those who had attended it, a request that the lectures should be published; this request was subsequently repeated in a letter from the Secretary of the Technical Committee. I therefore now give the lectures (with a few trifling exceptions) in the form in which they were delivered.

It did not appear possible in such a course, and if possible I should not have considered it advisable, to enter on any detailed criticism of all the various determinations of the heat equivalent; suffice it to say, the selection of examples was in no sense arbitrary.

Some reference to the educational aspects of the subject are, I hope, justified by the fact that the greater portion of my hearers were teachers of science.

It is natural that an audience should criticise its lecturer; but the reversal of the process may appear to be almost an impertinence. Nevertheless, I will venture to record some of the impressions I received during my visits to Yorkshire. Informal classes were held at the close of each lecture, and from the experience thus gained, as well as from a study of the large number of letters forwarded to me during the course, I was in some degree able to appreciate the keenness and ability which were the characteristics of those with whom it was my good fortune thus to come into contact.

The reflection that hundreds of such teachers should have been willing to sacrifice their Saturday afternoons to the study of certain physical measurements which did not even possess the charm of novelty, may somewhat lighten the gloomy prospect sketched for us by those who hold pessimistic views as to the future of Intermediate Scientific Education in this country.

I take this opportunity of thanking Mr F. H. Neville and Mr W. C. D. Whetham, not only for their assistance in the revision of the proof-sheets, but also for many valuable suggestions and criticisms.

E. H. GRIFFITHS.

SIDNEY SUSSEX COLLEGE,
CAMBRIDGE,
May, 1901.

CONTENTS.

LECTURE IV.

APPENDIX I.

APPENDIX II.

APPENDIX III.

ERRATUM.

p. 27, l. 5. *For* $\dfrac{981 \cdot 35}{987 \cdot 00}$, that is, $\dfrac{1 \cdot 0034}{1}$ *read* $\dfrac{981 \cdot 35}{978 \cdot 10}$, that is, $\dfrac{1 \cdot 0033}{1}$.

LECTURE I.

THE truth of the principle of the Conservation of Energy is now-a-days so firmly established, that there is some danger of students of natural knowledge regarding it as a doctrine to be accepted without question, rather than as a proposition capable of demonstration. It must be remembered that the proof, like that of all other natural laws, is a purely experimental one, and as all numerical relations dependent on experimental results are, at their best, but approximations, it is well to occasionally collect, and not only collect but also weigh, the evidence at our disposal.

It is possible that such an enquiry may to some appear as a matter of historical, rather than of immediate, interest. I shall, however, endeavour to show that much has been accomplished, even within the past few years, and that our progress in recent times has been real; although that progress partakes rather of the nature of an accurate survey of country whose main outlines are already familiar than of any advance into unknown territory.

It is not on that account, however, of the less consequence, for settlement is often as true an indication of progress as discovery itself.

It has been well said that the surest evidence of advance in any branch of knowledge is increase in the accuracy of our measurements of the phenomena involved, and, judged by this test, there is much to be proud of as we contemplate the knowledge of to-day and compare it with the somewhat indefinite condition of our thermal measurements some thirty, nay, twenty years ago.

In this course of lectures I shall endeavour to place before you our reasons for satisfaction, and my chief desire is to enlist your sympathies and your enthusiasm in the doctrine of the importance of accurate measurement and in the belief that, in physical science, accurate measurements are the steps by which we mount to heights otherwise inaccessible.

An illustration of the truth of this statement (if illustration is needed) is to be found in the history of the discovery of argon by Lord Rayleigh.

Who would have supposed it possible that his insistent endeavour to determine the density of nitrogen with the closest accuracy would lead to the discovery of a new element in the atmosphere?—a discovery which also led indirectly to the detection and isolation of other elements previously unknown.

As it happens, many of the results due to recent advances in physical science (such as the Röntgen rays, wireless telegraphy, &c.) have been, if I may so venture to describe them, of a sensational character, and there is, to my thinking, a danger that our younger scientists may, in their admiration for such achievements, neglect the more laborious and less attractive methods of research and measurement by which alone such discoveries are rendered possible.

To-day I am speaking to those who have responsibilities as teachers, and I trust that they will join with me in preaching, whenever possible, the doctrine that accurate measurements are necessary to physical salvation. Remember that he who accurately determines the value of any physical constant has placed a weapon of precision at the service of mankind.

One further remark by way of introduction. I propose, when possible, to illustrate what I have to say by means of simple experiments. The majority of these experiments require no complicated apparatus and could be performed with appliances to be found in almost every laboratory. Bearing in mind that this course of lectures is addressed to teachers, I have chosen such experiments *not* on account of their novelty or attractiveness, but by reason of their educational value and practicability. That the more simple an experimental illustration the greater its utility both to teacher and pupil, is a statement sufficiently near the truth to be accepted as a working hypothesis.

The discovery of the principle of the Conservation of Energy may be said to date from the time when Newton enunciated his third law. "Action and reaction are equal and opposite" is, in fact, the whole matter in a nutshell, although all the consequences involved in this statement were only fully comprehended and established in the latter half of the last century.

It is as Professor Tait has pointed out[1] a matter for regret that Newton's own explanation of the terms *action* and *reaction* has been so little considered and discussed by succeeding generations. He stated that there are two entirely distinct senses in which these words may be used, and that, whichever interpretation we accept, the law still holds true. Action in the one sense is a force only, and to this interpretation attention is, and has been, almost universally directed.

Newton's second interpretation of his third law is very different from this, and is of great importance. It is as follows:—

"If the activity of an agent be measured by the product of the force into its velocity and if similarly the counter activity of the resistance be measured by the velocities of its several parts, whether these arise from friction, adhesion, weight, or acceleration, &c., then activity and counter activity in all combinations of machines will be equal and opposite."

[1] *Recent Advances in Physical Science*, p. 33.

We must remember that by the *velocity* Newton meant the velocity in the direction in which the force is acting (*i.e.* the component velocity in the direction of the force).

A most important consequence of this interpretation is that the rate at which an agent does work is the rate at which the kinetic energy of a body increases. In other words the kinetic energy is increased by an amount equal to the work done in producing motion where the only resistance is that due to acceleration. Where the work is spent against friction, however, the visible energy of the system suffers decrease. If Newton had known of experimental evidence which indicated that the visible energy thus apparently destroyed was proportional to the heat developed against friction, it is possible that he might have enunciated the principle of the conservation of energy in its modern form[1].

---

[1] In the year 1606 (36 years before the birth of Newton) an anonymous author published a poem entitled "Hallo, my Fancy!"

You will find this remarkable production in Gilfillan's "Less-known British Poets," and the 13th verse runs as follows :—

"What multitude of notions
  Doth perturb my pate,
 Considering the motions
  How the Heavens are preserved,
    And this world served,
  In moisture, light, and heat !
If one spirit sits the outmost circle turning,
Or one turns another continuing in journeying,
If *rapid circles' motion be that which they call burning !*
  Hallo, my fancy, whither wilt thou go ? "

Here we have an attractive, although possibly unprofitable, subject for speculation. How far would the progress of science have been accelerated

The sure and rapid growth of our scientific knowledge in recent times is in a great measure due to the firm establishment of the principle of the conservation of energy. Had the truth of this principle been demonstrated by Newton, the progress of natural knowledge might have been quickened by a century!

Returning from such imaginings to sober reality, we find that from the time of Newton until the commencement (one might almost say the middle) of the nineteenth century, natural philosophers were heavily handicapped in their progress by the theory of caloric then firmly established. Theories of all kinds are useful scaffoldings; but he would be a bad architect who, instead of regarding the scaffolding as a temporary expedient, allowed its existence to cramp or influence the nature of his edifice.

The existence of an imponderable, indestructible fluid termed caloric was practically assumed in all discussions on natural phenomena, and many ingenious hypotheses and far-fetched explanations were put forward to account for the numerous difficulties which, in consequence, presented themselves.

For example, percussion was supposed to alter the condition of a body and lessen its capacity for heat. Thus in hammering a nail the caloric was simply squeezed out of the iron as the molecules were forced more closely together.

This explanation, however, did not appear to be

and what would have been our position to-day, had Newton but read that line "If rapid circles' motion be that which they call burning," and allowed his fancy to dwell thereon?

satisfactory in the case of lead, whose density is not increased by hammering.

When heat was developed by friction, part of the material was rubbed into powder and the Calorists insisted that the capacity for heat of the powder was smaller than that of the solid from which it was abraded. The absence of any experimental evidence of the truth of this hypothesis does not appear to have troubled them in any way! Such efforts to maintain the scaffolding unaltered necessarily hampered and retarded the efforts of the builders.

Count Rumford was the first to publicly question the popular caloric theory, when in 1798 he gave an account of his experiments. He placed a hollow gun-metal cylinder beneath a blunt steel borer and observed that after the cylinder had made about a thousand revolutions its temperature had risen from 60° to 130° F., while the scaly matter abraded by the friction weighed only 837 grains troy. "Is it possible," he writes, "that such a quantity of heat as would have caused 5 lbs. of ice-cold water to boil could have been furnished by so inconsiderable a quantity of metallic dust merely in consequence of a change in its capacity for heat[1]?"

The Calorists, however, were not convinced. Even when Rumford proved that the capacity for heat of the solid was the same as that of the dust, they said that, although the heat required to change the temperature of equal masses was the same, yet the solid metal contained a greater quantity of heat than the dust.

[1] *Rumford's Complete Works*, Vol. I. p. 478.

Rumford answered that if the heat were rubbed out of the material, a time must come when all its heat would be exhausted, whereas there was no evidence that such was the case. He also proceeded with further experiments in which the metal was immersed in water and, if we work out the results of those experiments, we find that 940 foot-pounds of work would raise one pound of water through 1° F.[1]. His final argument was as follows :—

" In reasoning on this subject we must not forget to consider that most remarkable circumstance, that the source of heat generated by friction in these experiments appeared evidently to be *inexhaustible.*

" It is hardly necessary to add, that anything which any *insulated* body, or system of bodies, can continue to furnish *without limitation* cannot possibly be a *material substance*[2]. It appears to me to be extremely difficult, if not quite impossible, to form any distinct idea of anything capable of being excited and communicated in the manner in which the heat was excited and communicated in these experiments, except it be motion." He adds, " I am very far from pretending to know how or by what means or mechanical contrivance that particular kind of motion in bodies which has been supposed to constitute heat is excited, continued, and propagated."

The historical importance of Rumford's experiments

[1] Rumford's experiments were qualitative, not quantitative. He made no attempt at the determination of any numerical relation.

[2] In an earlier portion of his paper (p. 479, *ibid.*) Rumford calls attention to the significant fact that the *rate* of production of heat remained constant so long as the work was done uniformly.

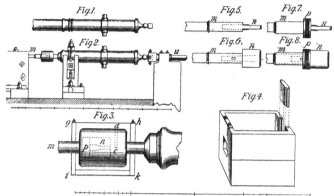

Fig. 1.

"Fig. 1 shows the cannon used in the foregoing experiments in the state it was in when it came from the foundry.

Fig. 2 shows the machinery used in the experiments No. 1 and No. 2. The cannon is seen fixed in the machine used for boring cannon. *u* is a strong iron bar (which, to save room in the drawing, is represented as broken off), which bar, being united with machinery (not expressed in the figure), that is carried round by horses, causes the cannon to turn round its axis.

   m is a strong iron bar, to the end of which the blunt borer is fixed; which, by being forced against the bottom of the bore of the short hollow cylinder that remains connected by a small cylindrical neck to the end of the cannon, is used in generating heat by friction.

Fig. 3 shows, on an enlarged scale, the same hollow cylinder that is represented on a smaller scale in the foregoing figure. It is here seen connected with the wooden box (*g*, *h*, *i*, *k*), used in the experiments No. 3 and No. 4, when this hollow cylinder was immersed in water.

   p, which is marked by dotted lines, is the piston which closed the end of the bore of the cylinder.

   n is the blunt borer seen sidewise.

   d, *e* is the small hole by which the thermometer was introduced that was used for ascertaining the heat of the cylinder.

Fig. 4 is a perspective view of the wooden box, a section of which is seen in the foregoing figure. (See *g*, *h*, *i*, *k*, Fig. 3.)

Figs. 5 and 6 represent the blunt borer *n*, joined to the iron bar *m*, to which it was fastened.

Figs. 7 and 8 represent the same borer, with its iron bar, together with the piston, which, in the experiments No. 2 and No. 3, was used to close the mouth of the hollow cylinder."

is so great that it may interest you to see this copy of his original drawings. The description of the figures is extracted from his paper entitled "An Inquiry concerning the Source of the Heat excited by Friction[1]."

Rumford's work was of the highest value. As Professor Tait remarks[2], it was throughout free from that *a priori* style of reasoning which had hitherto been so fatal to the progress of natural science. Had Rumford shown that the heat developed by the solution of a powder in acid was equal to that developed by the solution of the same mass of the solid, he could have claimed the sole credit of having established the doctrine of the non-materiality of heat[3]. Nevertheless Rumford's conclusions remained for another forty years as subjects for ridicule rather than for admiration.

Almost immediately after the publication of Rumford's paper, Sir Humphry Davy proved experimentally that two pieces of ice may be melted by rubbing them together, and thus he gave conclusive proof (although there is evidence that he himself at the time did not perceive it) that heat is not a form of matter, and therefore his experiments are historically of the first importance. In a second series of experiments he so

[1] *Rumford's Complete Works*, Vol. I. p. 492.

[2] *Recent Advances in Physical Science*, p. 42.

[3] This statement, however, requires some qualification, as it is not certain that the heat developed in both cases would be the same. For example, the heat developed by the solution of a steel spring when wound up would differ from that developed by the solution of the same spring when not in a state of strain. In other words, it is not sufficient that the masses should be equal, their physical conditions must also be identical.

contrived that the friction between the lumps of ice took place in the exhausted receiver of an air-pump. He says :—

" From this experiment it is evident that ice by friction is converted into water, and according to the supposition its capacity is diminished ; but it is a well-known fact that the capacity of water for heat is much greater than that of ice ; and ice must have an absolute quantity of heat added to it before it can be converted into water. Friction, consequently, does not diminish the capacities of bodies for heat."

It was not, however, till 1812 that he enunciated this proposition : " The immediate cause of the phenomenon of heat then is motion, and the laws of its communication are precisely the same as the laws of the communication of motion[1] "; and, on reflection, it seems extraordinary that the publication of the works of Rumford and Davy produced so little effect, and that their conclusions should have been regarded merely as ingenious hypotheses until the time of Joule.

It would be impossible in any historical summary, however brief, to omit the name of Dr Julius Mayer (1842), as he was the first to employ the phrase, *the mechanical equivalent of heat*, and thus enunciate in a distinct form the law of the Conservation of Energy[2].

It is doubtful, however, if Mayer has deserved all the credit which was in consequence at one time assigned to him. The data on which his conclusions were based were

---

[1] *Elements of Chemical Philosophy*, 1812, pp. 94, 95.
[2] *Ann. Ch. Pharm.*, XLII. 233.

not sufficient, while certain of his assumptions were undoubtedly erroneous. Mayer denied on more than one occasion that heat depends on motion, and yet he has been called by some "the discoverer of the modern theory of heat." Nevertheless he, as it happens, rendered a service to the cause of science by distinctly, although perhaps over-boldly, enunciating a theorem of the highest importance. In this case no evil consequences resulted from the premature publication, for the work of Joule and of Colding (already partially accomplished) afforded satisfactory and complete experimental evidence of the truth of the proposition advanced by Mayer.

Let me here briefly summarise the condition of our knowledge at the time of the advent of Joule.

Although not generally apprehended, the principle of the Conservation of Energy had, in its main features, been enunciated by Newton. Rumford and Davy had demonstrated that heat is not matter; but their conclusion was not generally accepted. The school of Calorists still existed, and although the general principle of Conservation of Energy had been enunciated by Mayer, his data were so scanty, and often erroneous, that his conclusions carried but little conviction to men of science. What the situation demanded was rigorous experimental proof that apparent disappearance of energy, when work was done against friction, was invariably accompanied by the generation of a proportional amount of heat, and that this quantity of heat was independent of the manner in which the work was done, or the nature

of the materials employed. Hence the singular importance of Joule's experiments.

The actual numerical values obtained by him are, we
now know, somewhat inaccurate, but, at the same time,
marvellously near the truth when we consider the conditions under which he worked, and more especially the
unsatisfactory state of thermometry in his time.

It is not so much the *accuracy* as the *variety* of the
evidence supplied by Joule which claims our admiration.
He did not content himself with the examination of some
single case of transformation of energy. He investigated
the heat caused by the friction between solids, and also
that developed in overcoming fluid resistance, not only in
the case of one fluid but of different fluids.

Again, the *manner* of doing work was varied from the
descent of weights to that performed by an electric current
in overcoming resistance. Although many stones have
been laid by other workers it may be said, I think without
exaggeration, that the piers were built by Newton and
the keystone of the arch supplied by Joule.

The principle of the Conservation of Energy states
that in all its forms energy remains a constant quantity,
however many transformations it undergoes. In other
words, if energy is made to pass from any condition such
as that of matter in motion into any other condition such
as molecular, or electrical energy, the numerical value of
the resulting effect depends simply on the quantity of
energy so transformed, not on the method of transformation, the materials, time, or any external conditions.

All natural phenomena have their origin in the transference of energy from one form to another, and energy is only available when it is capable of transformation. The total energy of the molecules of air in this hall is, no doubt, very great; but we have not the means of readily translating that energy into other forms and hence can make no practical use thereof. When we refer to the "available energy of a system" we mean that portion of it which we *can* utilise.

It must be remembered that all our measurements are but relative; when we speak of a mass at rest we mean that its motion is that of the surface of the earth at the same place. We then say that it has no kinetic energy, whereas, in reality, all matter that we know of is in rapid motion and the kinetic energy of any mass which appears to us to be stationary is in reality very great[1]. It is well therefore to keep this relative nature of our measurements in mind, although when, in the future, I speak of the kinetic energy of a system it will be understood that I refer the motion of its parts to that of the surface of the earth at the same place.

I now wish to draw your attention to the more

[1] For example, take a lump of our best coal and give it a velocity of about 5 miles per second; its energy due to *movement* would then be about the heat equivalent of its complete combustion. Now, if we regard the sun as fixed, the velocity of the earth in its orbit is between 19 and 20 miles per second. Hence the energy of a mass of coal due to its velocity (although it appears to us to be at rest) is far greater than its chemical energy. In other words, suppose the earth to be constructed entirely of the best coal, the heat generated by the sudden stoppage of its movement would be about 15 times as great as the heat which would be developed by its complete combustion!

common phenomena of the transformation of energy. The examples which we shall consider are probably known by all of you; but their importance is so great that it is a tale worth telling, although twice told, and I would ask you more especially to fix your attention on the fact that the cases we consider present one characteristic feature, viz. that when we convert a given quantity of energy from the form A to the form B some portion of it is, at the same time, converted into heat.

Experiment. This arrangement of a rotating tube filled with water is probably familiar to most of you. You see that as the tube is rotated great heat is developed at the place where it is clasped by the wooden pincers, so that the contained water is boiled and the cork explosively expelled. Let us suppose that but 1 grm. of water was here raised to boiling point and converted into steam, then the work which my assistant has done was sufficient to raise a weight of about 640 grms. (*i.e.* about 1·4 lbs.) from the bottom of a coal-mine 1,400 ft. deep.

Simple as this experiment is, it is important educationally if we show that the result is the same, no matter what the material of the pincers or of the tube, provided that the work done against friction remains constant. Here then we have the familiar case of conversion of work directly into heat. The source of the energy is to be found in the chemical separation of the substances eaten by the operator.

Experiment. Here, again, you see that by hammering the lead its temperature is sufficiently raised for it to ignite the phosphorus I place upon it. This is an example of the conversion of kinetic energy into heat.

Experiment. By suddenly compressing the air in this tube, the heat generated is again sufficient to raise the temperature to the ignition point of phosphorus.

A neat application of this method of transforming energy comes to us from the other side of the Atlantic. When driving piles they make the upper portion of the pile hollow, and a heavy cylinder, which just fits the hollow tube, is supported above the pile, while a little gunpowder is placed at the bottom of the hollow. When the cylinder is released, it compresses the air within the hollow of the pile so rapidly that the heat generated explodes the gunpowder and the force of the explosion drives the cylinder up to its original position (where it is caught), while the reaction drives the pile further into the ground.

Experiment. Here I have a bottle of carbon-dioxide, the pressure within the cylinder being at least 60 or 70 atmospheres. As we open the tap the contents stream out at the nozzle, there is rapid evaporation, a large volume of air under atmospheric pressure is displaced, and thus mechanical work is done. Heat, therefore, disappears at such a rate that the temperature of the emergent substance is reduced below $- 80°$ and solidification takes place. You see the resulting carbon-dioxide snow, and when I add some of it to this mercury, you perceive that I am able to lift the frozen lump of mercury by the wire placed in it. This is, of course, a converse case to the last, where we raised the temperature of a gas by compression.

In considering this experiment it is important to remember that the heat which has thus disappeared corresponds to that developed by the work done when the cylinder was originally charged. If the vessel had been an adiabatic or perfectly non-conducting one, and had therefore retained its contents until to-day at the temperature to which they were raised at the time of charging, the sole effect of now allowing the contents to escape would have been to lower them to practically their original temperature. Hence, on the whole, we have had no real disappearance of heat; in fact, could we minutely trace all the steps of the cycle, we should find that a certain quantity of energy had been degraded. By the whole cycle of operations we have therefore increased rather than diminished the earth's store of heat.

Experiment. Another form in which energy presents itself is in the separation of electrically charged bodies. If I inductively charge this electroscope and then remove the negatively electrified ebonite rod, I have to overcome the attraction of the positively charged electroscope. Work is thus done upon the system in consequence of its electrical condition, as is shown by the divergence of the gold leaves whose centres of gravity are thus raised. When I connect the electroscope with earth, that energy is lost and, could we examine with sufficient accuracy the temperature of the wire connecting it with earth, we should find that it had been increased. When a body at a higher potential is discharged, the heat generated is usually rendered evident by the familiar phenomenon of the spark. Thus we continually find that as we alter the energy of an electric system, some of that energy appears as heat.

The conversion of the energy of an electric current into heat is now-a-days so familiar a phenomenon that it is almost unnecessary to call your attention to it. All our electric lighting stations are engaged in converting some of the heat generated, by the union of the carbon and hydrogen of coal with the oxygen of the atmosphere, into mechanical energy; this in its turn is changed into the energy of an electric current, which is again degraded into heat in our incandescent lamps, a very small percentage of the total energy appearing as visible radiant energy. The electric current here is merely the most convenient method of converting mechanical energy at one place into heat and some visible radiant energy at another.

Experiment. I have here a fly-wheel so connected with a number of magnets that the latter rotate with great rapidity as the fly-wheel revolves. You will see that the resistance to movement is small and that the fly-wheel continues to rotate for some time after the hand has been removed. We could form an idea of the energy of this rotating system by finding the distance through which it would raise a known weight.

The magnets are revolving within the coils of an

incomplete electrically conducting circuit but without touching them. We will now cause the fly-wheel to rotate at, as nearly as possible, the same rate as before, and then join the ends of the circuit and thus allow the induced currents, formed by the rotation of the magnetic field, to pass through the coils. These currents will always be established in such a direction as to resist the motion of the magnets. Hence more work is now done per revolution by the revolving fly-wheel than before I closed the circuit and thus the wheel is brought to rest as if by the application of a powerful brake. As a consequence, the energy of the fly-wheel disappears after a very few revolutions and at the same time you see by the glowing of this wire that heat has been developed in the electric circuit, this heat being the equivalent of the energy which, in the previous case, enabled the fly-wheel to perform many more revolutions[1].

I particularly wish to draw your attention to the fact that the manner in which that work is being done is of no importance. The heat generated will be the same, whatever the cause of the stoppage, if it can be shown that the energy of the rotating wheel has not appeared in other forms. It is this independence of the nature of the material, or the manner in which the resistance is overcome, which is the true test of the principle of the conservation of energy.

Experiment. In a cup formed in the top of this

[1] When the electric circuit was incomplete the fly-wheel performed 27 revolutions; when the circuit was closed it came to rest after 5 revolutions.

copper cylinder, which is rotating in a strong magnetic field, I place a piece of phosphorus. I think you will see that, in consequence of the resistance offered by the currents due to its rotation in the magnetic field, the work done on the copper is sufficient to raise its temperature sufficiently to ignite the phosphorus.

Sound-waves are, themselves, carriers of energy and, as is well known, a certain amount of heat is developed by the production of sound. The dissipation of energy in the sound-wave itself is, however, very small, otherwise the waves would die away with extreme rapidity. It is, nevertheless, certain that when sound-waves impinge on solid bodies a certain amount of heat is developed.

Experiment. Here is an example of such a propagation of energy. I set this tuning-fork in vibration and, when I stop it, I think you will distinctly hear the same note given out by the second fork.

Again, by suspending a pith-ball against one prong of the receiving fork, we can (with the aid of the lantern) render visible to the eye the vibrations set up by means of the energy transmitted from the first fork to the second. I have not been able to devise any method of showing you that heat *has* in this case been actually generated in the receiver, but as you can see the rapid hammering of the metal by the ball, it is easy to understand that such impacts must be accompanied by the evolution of some heat.

Of all forms of energy we, on this earth, are most indebted to the radiant form; the source alike both of the energy of the food we eat and of the coal we burn, it is at once, the greatest and the highest of the forms in which energy presents itself. We have few machines which work by the direct transmutation of radiant energy to lower forms, yet we must remember that the action of all water-mills, turbines, &c. is dependent on the transformation of radiant into potential energy—a transformation performed for us by nature, without our assistance. As our coal supply becomes exhausted I think it possible that engineers will have to turn their attention more seriously to the direct conversion of the energy so freely supplied to us by the sun.

You know that power is now transmitted from the Falls of Niagara to towns at considerable distances, and here we have an example of an indirect use of radiant energy. As the solar radiant energy reaches our earth it is almost entirely degraded to the heat form. A portion of this heat is transformed, by means of evaporation, into potential energy of masses of water at a high elevation, and some of this energy is retained by the water after it has fallen on the American Continent, since it is then at a higher level than the ocean from which it sprang. During its further descent at Niagara this water is so directed that it causes the rotation of conducting bodies in a magnetic field; thus a portion of its energy is converted into the energy of electricity in motion and, in this transformation also, a portion is degraded into heat.

Again, the energy of this electric current is, at distant places, almost wholly converted into heat, either by the use of incandescent lamps, or by overcoming the friction

of various pieces of machinery. Here we have an
interesting series of transmutations; but I would press
on your notice the fact that no single one of them has
been accomplished without the payment of a forfeit in the
shape of a certain per-centage of the total energy degraded into the form of heat.

How small a fraction of the earth's share of the solar
energy is directly utilised by man is evident from the
following considerations.

According to Lord Kelvin's calculations the energy of
the sun's radiation is equal to about 7000 H.P. per square
foot of its surface.

Some observations (by J. Y. Buchanan[1]) taken in Egypt
with an improved form of calorimeter, lead to the conclusion that each square metre of the earth's surface,
which is exposed perpendicularly to the sun's rays, receives radiant energy equivalent to 1 H.P.

Now, the area of the great circle of the earth is
roughly 130×10^{12} square metres; thus the working value
of the sun's radiation to us is about 130 billion H.P. If
we deduce from this the H.P. radiation per 1 square foot
of the sun's surface, we obtain a value very decidedly less
than that obtained by Lord Kelvin. It is probable, therefore, that the result obtained by Buchanan does not err
on the side of excess.

I am aware that such figures as a billion convey little
meaning to our minds. Recently, I came across an
example which illustrated in a somewhat novel manner the
vastness of this number. Do you suppose that there are

[1] *Proc. Camb. Phil. Soc.*, 1900.

a billion bricks in all London? I admit that I should have considered it possible. Now, assuming the area of London to be 100 square miles, simple arithmetic will tell you that 1 billion of ordinary sized bricks placed together would cover London in a solid mass to the depth of 25 feet. Remembering this, I ask you to reflect on the magnitude of the stream of energy thus continuously poured upon the earth;—130 billion H.P.!

I find that the latest estimate of the total population of the earth is 1500 millions. This would give $\dfrac{130 \times 10^{12}}{15 \times 10^{8}}$ H.P. per individual. In other words, if the radiant solar energy, falling on the earth were wholly converted into mechanical energy, each individual's share would enable him to lift a weight of 33,000 lbs. through a vertical distance of 100,000 feet (nearly 20 miles) every *minute* of his life.

Experiment. In this familiar apparatus (Crooke's Radiometer) you see apparently a direct transformation of radiant energy into mechanical motion; in reality, the change has taken place in two steps. The radiant energy is changed into the heat form (more being thus converted on the black, than on the bright, side of the vanes) and a portion of the heat thus generated increases the rapidity of the motion of the molecules of the gas there present, thus setting up, on opposite sides of the vanes, a difference of pressure sufficient to cause rotation; but this kinetic energy again becomes converted into heat by the friction of the bearings.

The form of energy chiefly used by the engineer

is that of chemical separation. The burning of coal is the conversion of such energy into the heat form, the heat being probably of kinetic origin and due to the impact of atoms. We must bear in mind, however, that in such cases we always have a breaking up of existing molecules, as well as the formation of new ones.

When we oxidise zinc by the action of dilute sulphuric acid, we start with molecules of zinc, sulphuric acid, and water, and we finish with zinc sulphate, hydrogen, and water. The chemical energy of the products will be less than that of the original substances and the difference is accounted for by the heat developed by the reaction. If we connect the zinc externally with a plate of some other metal immersed in the acid we obtain an electric current, and by means of this current we can perform mechanical, or other, work in which case less of the lost energy of the compounds will appear directly as heat.

Experiment. I cause this current to decompose water and in consequence less heat will (for a given consumption of zinc) be developed in the battery and circuit than would be the case if I merely connected the ends of the battery wires. The energy thus again appears as that of chemical separation, although some portion has certainly gone downhill as heat during the process. I apply a light to the soap-bubbles which contain the separated gases. You all know the exceedingly sharp explosion thus obtained. Here nearly all the apparently missing heat reappears at the place of explosion, although a fraction may be generated elsewhere by means of sound-waves.

To-day we have concerned ourselves chiefly with the transference of various forms of energy into heat and we have found that, in nearly all cases, such conversion is easy of accomplishment. We have seen that the change from any form *A* to another form *B* involves the appearance of *some* of the total energy as heat. Each time we alter our investment in energy, we have thus to pay a commission, and in future lectures I hope to be able to show you that the tribute thus exacted can never be wholly recovered by us and must be regarded, *not* as destroyed, but as thrown on the waste-heap of the Universe.

In conclusion, it is necessary to remember that, although the doctrine of the indestructibility of energy is established beyond all question, it in no way involves the assumption that energy in different forms is equally available.

LECTURE II.

IN my last lecture I gave a very brief outline of the early
history and the general principle of the Conservation of
Energy. I propose to-day to direct your attention more
particularly to the nature of the measurements which
have to be made before we can obtain the numerical
relation between the mechanical, and the thermal, forms
of energy.

Let us first consider the measurement of the potential
and kinetic energy of a system.

The most common English method of stating potential
energy is in *foot-pounds*. This is unfortunate, as we
require further information (viz. the local value of g)

before we can attach any definite value to an expression of this kind.

The ratio of the value of 1 ft.-lb. of work here (or rather in Manchester) to its value at the Equator is $\frac{981\cdot35}{987\cdot00}$, that is, $\frac{1\cdot0034}{1}$, and this ratio increases in magnitude as we approach the Poles. If, therefore, the potential energy of a system is given in ft.-lbs., and the position of the observer is not stated, there exists an uncertainty of about 4 parts in 1000, and even when the position is known, a troublesome reduction to a common standard is necessary. It is far better to adopt at once a non-gravitational and absolute unit, such as the " erg," or dyne-centimetre.

Now a mass of 1 gramme, if acted on by its own weight, would in this room move with an acceleration of 981·4 cm. per sec. per sec. Hence, as I support this 1 gramme weight in my hand, I must be giving it an equal upward acceleration; that is, I am exerting upon it an upward force of 981·4 dynes. Suppose therefore that I lift a mass of 100 grammes through a vertical distance of 101·9 cm., I have caused a force of 98140 dynes to move through a distance of 101·9 cm.; hence I have done almost exactly 10^7 ergs, that is, 1 *joule* of work. It may be convenient to remember this example, viz. that a *joule* of work is done (approximately) whenever we lift 100 grammes through a vertical distance of 102 cm., the exact distance depending, of course, on the local value of g; also, if I perform this work in one second, then the power (or rate of doing work) is 1 *watt*.

I propose to adopt the erg, or its multiple the *joule*, as our unit in all cases, and to express the results obtained by different observers in terms of that unit[1].

The *potential energy* of a system may be defined as the work it can do in consequence of its position or configuration; the *kinetic energy* of a system is that due to motion.

We know from a study of the phenomena of acceleration that if a mass m is shot upward with a velocity v, it will rise through a vertical distance s such that $s = \dfrac{v^2}{2g}$.

Now, this mass is pulled towards the earth with a force mg; hence if allowed to descend through the vertical distance s it could do $mg \cdot s$ units of work. By its ascent to its highest position, therefore, its potential energy must have been increased by this quantity, $mg \cdot s$. This increase was entirely due to its motion when starting upwards. Hence the kinetic energy of a mass m moving with a velocity v is equivalent to the potential energy

$$mg \cdot s = mg \cdot \frac{v^2}{2g} = \frac{1}{2} mv^2.$$

Thus since $mg = f$ dynes (in the C.G.S. system) we know that $\frac{1}{2}mv^2 = fs$ ergs.

---

[1] The conversion from ergs to foot-pounds is always a troublesome one and involves a knowledge of g at the place of observation. As it is often necessary to convert the mechanical equivalent expressed in ergs to foot-pounds at Greenwich ($g = 981 \cdot 24$) the following factors may be found useful.

Mechanical equivalent in ergs $\times \cdot 0000334363 = mechanical\ equivalent$ expressed in foot-pounds at Greenwich.

Mechanical equivalent in foot-pounds at Greenwich $\times 29907 \cdot 6 = mechanical\ equivalent$ expressed in ergs.

It is important to notice that $\frac{1}{2}mv^2$ is not a directive or vector quantity; it can never be negative. Hence the kinetic energy of a system is the sum of the kinetic energy of its parts, independently of the directions in which those parts are moving, although the availability (or what Lord Kelvin terms the *motivity*) may depend upon the direction of such movements. The energy of a swarm of gnats is the same whether they are all moving in different directions, or all in the same direction. In the first case we may not, but in the second case we may, be able to utilise that energy.

We are now, I think, in a position to define the *mechanical equivalent* and I propose to do so as follows:—

" The *mechanical equivalent* is the number of ergs which, if wholly converted into heat, would generate one thermal unit."

In order, therefore, to obtain the value of this constant, we have two distinct sets of measurements to perform.

(*a*) The accurate measurement in ergs of the change in the mechanical energy of a given system ;

(*b*) The measurement of the quantity of heat generated by the complete conversion of that number of ergs of energy into the form of heat ; and I may add that it is the second of these measurements which has presented the greatest difficulties.

I remember, some years ago when examining in the Cambridge Local Examinations, I asked for a definition of the *mechanical equivalent*. I received many curious

answers; but one, in particular, impressed itself upon my
memory, viz., " The object of the *mechanical equivalent*
is to waste as much work as possible." Now, the most
lenient of examiners could scarcely have endowed this
answer liberally with marks; at the same time it has often
struck me that there is a good deal of truth in it. If we
start on a determination of this important constant, our
object is to transform the *whole* of the energy we measure
into heat, and that heat all generated at one place. We
do want to " waste as much work as possible."

If we have a mass at a high elevation, then, if we know
the value of g, we have seen that we can calculate with
accuracy its available energy, that is, the number of ergs
of work it can do in descending through a given distance.
Suppose, in its descent, it turns a paddle-wheel ; some of
the energy we measured will have been dissipated by
sound-waves, some by friction at bearings exterior to the
calorimeter, some perhaps by the mass having a certain
amount of kinetic energy at the end of its descent. The
difficulty, in fact, is not so much that of " wasting all the
work " as of wasting it at the place where we measure its
equivalent.

The various devices by which these difficulties have
been overcome will be considered in my next lecture.
To-day I more especially wish to fix your attention on the
method by which we measure the heat generated, on the
assumption that we are able to convert all the energy of
the system into heat. Two distinct methods of measuring
quantities of heat suggest themselves :

(1)   Determination of the quantity of heat required

to change the physical condition of a body (to melt, for example, a given mass of ice), or,

(2) Observation of the heat required to raise a certain mass of some selected substance, such as water, through a given range of temperature.

The first of these methods possesses one great advantage, it is independent of all temperature measurements.

Unfortunately there are practical difficulties. For example:—The magnitude of the unit thus obtained is, in many respects, inconvenient, and we do not as yet know a satisfactory method of determining its value with accuracy. It has also been recently proved that the density of ice is variable[1].

Nevertheless, so great is the advantage due to the absence of thermometric measurements, that some of our leading authorities are still in favour of the adoption of such a unit. (See Appendix I.)

The second method of measuring heat (by the rise in temperature of a known mass of water) is, however, the one which has been most universally adopted from the time that calorimetry became a science. And, in this matter, I suppose we must yield to custom. The choice is an unfortunate one, for not only is the accurate measurement of temperature one of the most difficult of all measurements, but also the material selected, viz. water, is capricious in its behaviour.

It must be remembered, however, that the true primary unit is the heat-equivalent of 1 erg.

[1] *Physical Review*, VIII. 1899, pp. 21—38.

We must therefore regard such units as secondary ones, chosen simply for convenience; just as, for example, the ohm and the volt are arbitrary multiples of the c.g.s. units of resistance and potential difference, so the thermal unit, as ordinarily defined (viz. the quantity of heat required to raise 1 gramme of water through 1° at a certain temperature) is some multiple of the heat-equivalent of one erg.

The ratio of the ohm, or the volt, to the corresponding c.g.s. unit is, however, unaffected by the properties of any material substance; whereas the ratio of this secondary thermal unit to the primary one is dependent upon the behaviour of a particular compound, and this dependence has been a cause of difficulty and confusion. There is, I am afraid, little hope of any change; nevertheless I am convinced that the only satisfactory method would be to wipe the slate clean and start afresh. Let our secondary unit of heat be the *joule*, or 10^7 ergs; we should then give the capacity for heat and latent heat of fusion etc. of a substance as so many *joules*. This alteration in nomenclature would certainly tend to simplification, especially in the study of steam-engines and other practical applications of thermodynamics.

In the meantime we must take things as we find them (or rather as the majority insists on regarding them), and, therefore, try to determine the real value of this secondary *watery* unit: in other words, how many ergs will raise 1 grm. of water through 1° at a certain temperature?

Here, of course, comes the question—what is one

degree of temperature? If the whole of this course of lectures was devoted to the reply the time would be all too short. Let us put the question in a more practical form and ask "How do we form a scale of temperature?" I would answer as follows :—

We observe some property of a body (such as length, pressure, electrical resistance, etc.) which undergoes alteration with change of temperature. We then find the condition of the body at two fixed, or recoverable, temperatures, and we make the assumption that at other temperatures the change in the body is proportional to the change in temperature.

Now, is there any case in which this assumption is true? If so, how are we to prove it?

There is one test we can apply, which will tell us, at all events, that the various scales thus obtained do not agree and that, therefore, no two of them can be correct. We have one fundamental definition of equality of temperature about which there can be no doubt, viz. if *A* and *B* are at equal temperatures, then if *A* and *B* are placed in contact their temperatures will not alter.

Suppose you observe the position of the mercury in a glass thermometer, first in melting ice, and then in steam at standard pressure. Make a mark half-way between these two points. Now place this thermometer in some water and change its temperature until the mercury stands at this half-way mark. Repeat these operations, using some other kind of matter, say a gas, as the thermometric substance; you will obtain two lots of water whose temperatures are presumably equal, for, according to the scales thus formed, they are each half-way between the temperatures of melting ice and steam. Mix these two masses of water; in this case, where a gas has been used as the second material, you will find that the temperature of the first mass rises, while that of the second falls! How, therefore, can we find out which (if any) of our various property-of-matter scales are true? What is the ideal scale to which all must be referred?

It is one of the many claims which Lord Kelvin has upon our gratitude that he was a leader amongst those who furnished us with this ideal, or absolute scale. Most of you are probably acquainted with that imaginary, but nevertheless exceedingly useful piece of mental apparatus named Carnot's engine. Although this fairy engine was built of impossible substances and based on wrong conceptions as to the nature of heat, I doubt if many of our existing engines have done more useful work, and I would earnestly impress upon you the necessity, if you wish to obtain a real grasp of thermodynamics, of studying Carnot's cycle until you feel that you have fully mastered it.

Time will not permit us to fully enter into this matter; but I ask the forbearance of those who are already well acquainted with this portion of our subject while I briefly describe the nature and application of Carnot's engine; especially as I propose to present it to you in a somewhat different form from that usually given in the text-books.

In an ordinary steam-engine we are confronted by a complex problem; details as to the nature of steam, construction of valves, conductivities of substances, etc.

distract the attention from the real essence of the matter. Now, in Carnot's engine we get rid of all this, just as in geometry we reason on the properties of straight lines, without bothering our heads as to the effect of the irregularities which in practice invariably present themselves.

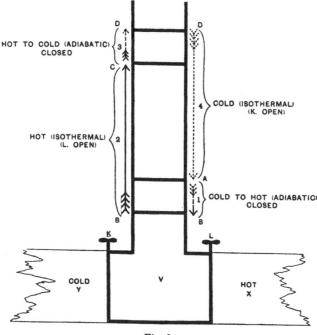

Fig. 2.

Suppose the walls of this central vessel (V) to be constructed of a perfectly non-conducting (or adiabatic) substance. Let the sides of this vessel come into contact, on the right hand, with a perfectly conducting hot body

3—2

X, on the left with a perfectly conducting cold body Y. Let us, however, have the power of sliding away, by means of the handles L and K, the adiabatic faces which prevent any passage of heat to or from X and Y. It must be understood that no work is done in removing or replacing these adiabatic shutters, we simply issue a command and they are gone. Further, they are shutters whose removal allows the passage of heat only; not of any *material* which is within the vessel V.

Thus, for example, remove K. Now the gas, or other material within the vessel, is brought into thermal contact with the cold body and we will suppose that it at once assumes the same temperature; but when K is replaced the contents of V are uninfluenced by the presence of Y. In the same manner, remove the shutter at L, the contents are at once brought into contact with and assume the temperature of the perfectly conducting hot body X; but in neither operation does any *material* leave, or enter, V.

A tube (also formed of the adiabatic substance) is fitted into the upper part of V and closed by an adiabatic piston. Suppose that we find K open and L shut, so that we know that the contents of V are at the temperature of the cold body; close K and suppose that the piston then stands at A. Now commence your cycle. Force the piston downwards; work is thus done *on* the contents and the heat produced by this work will cause the temperature within the vessel to rise. Let this continue until the temperature within V is that of the hot body and let the piston then be at B. Now, slide L away, and then allow the piston to rise; work is done *by* the contents and heat disappears, although the temperature is kept constant by the transference of heat from *X* to *V*. Let this stroke terminate at any position you please, say at *C*; then close *L*, and still further diminish the external pressure. As the piston again rises, heat again disappears, but in this case the temperature will fall. Allow this process to continue until the contents arrive at the temperature of the cold body. Let this happen when the piston is at *D*. Now slide *K* away and force the piston downwards until it is at its starting point *A*. Although work is thus done on the system, the temperature of *V* is not increased, as the heat generated passes into *Y*. Finally close *K*. We have now completed a cycle, for the material in *V* is exactly in the same condition as regards quantity, temperature, pressure, and volume as when we started. The total effect has been the removal of heat from the hot body and the addition of some heat to the cold one, and that without exposing the working substance in *V* to contact with bodies at a different temperature from its own.

Let us consider the four steps.

(i)   *A* to *B* ↓ contents cold to hot, adiabatic.
(ii) *B* to *C* ↑   „   hot   isothermal.
(iii) *C* to *D* ↑   „   hot to cold, adiabatic.
(iv) *D* to *A* ↓   „   cold   isothermal.

From this summary it is evident that the average temperature (and therefore pressure) during the two upward strokes (2nd and 3rd) is greater than during the two downward ones (1st and 4th). Now as the lengths of the total upward and downward movements are the same, it follows that the total work done during the rising strokes exceeds that done during the falling ones. Thus work has been done *by* the system, while heat has been taken from X and some heat passed into Y.

If the working substance in V is a gas the relation between the volume and pressure of the contents, at each stage of the operations, can be represented diagrammatically as in Fig. 3. For example, the ordinates at A, B, C and D are proportional to the pressures within V when the piston was in the positions indicated by the same letters in Fig. 2.

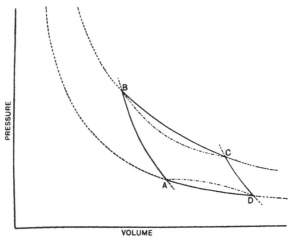

Fig. 3.

The quadrilateral figure $ABCD$[1] is bounded above and below by isothermal, and at its sides by adiabatic, curves;

---

[1] The meaning of the dotted lines *within* the quadrilateral will be explained in a subsequent lecture, for our present purposes they may be disregarded.

the enclosed area represents the excess of the work done
during the upward, over that done during the downward
strokes.

Now, if you consider the operations, you will find that,
by commencing with the piston at C (Fig. 2) and the
contents at the temperature of the hot body, it is possible
to complete the cycle by taking all the steps in reverse
order[1]. In this case, heat will be passed into the hot,
while some heat will be taken from, the cold body. Hence
the average pressure during the downward, will be greater
than that during the upward, strokes and work is done *on*,
instead of *by*, the working substance.

If you construct the corresponding diagram, you will
discover that the resulting quadrilateral figure is the same
as that obtained from the first or forward cycle. When
sketching it, however, your pencil will pass through any
point on its perimeter in the opposite direction to that
previously followed. Thus work has been done *on*, or *by*,
the working substance, according as we have travelled
round the work-space in the same, or the opposite,
direction to the hands of a clock.

Carnot, at the time he wrote his paper, considered
that this transference of heat from a high to a low
temperature was analogous to the falling of a given mass
of water from a high to a low elevation, and that the work
done in the transfer was not due to any consumption of

---

[1] The operations are as follows :

(1)	C to B ↓	L open	contents hot.		
(2)	B to A ↑	shutters closed	,,	hot to cold.	
(3)	A to D ↑	K open	,,	cold.	
(4)	D to C ↓	shutters closed	,,	cold to hot.	

the substance called caloric, but was simply performed by it when sliding down a temperature gradient. We now know that the heat delivered to Y is less when the cycle is completed than that taken from X, and the difference is the heat-equivalent of the work done during the cycle. Also, when we do work on the system during the reverse cycle more heat will be given to X than was taken from Y, and if the ratio $\dfrac{\text{Heat added to } X}{\text{Heat taken from } Y}$ during a reversed cycle has the same value as the ratio

$$\frac{\text{Heat taken from } X}{\text{Heat added to } Y} \text{ during the direct operation,}$$

we know that the engine is a perfect one and that it has the highest possible theoretical efficiency.

This is a very important point in our argument, so let us consider it more fully. Suppose it possible to obtain a better engine, better in the sense that it converts into work during a forward cycle a larger proportion of the heat received from X than is thus converted by the reversible engine.

Now let this better engine be so coupled up with the reversible one that it works it backward, stroke for stroke. The reversible engine will each time remove a larger quantity of heat from Y during its backward cycle than the more efficient engine puts into Y during a forward one. Thus the quantity of heat in the hot body can remain constant while the quantity in the cold must steadily diminish. True, the compound engine thus formed will, on the whole, do work; but only by the transference of heat from a colder to a hotter body. Now, this is contrary to all experimental evidence, for we have overwhelming proof that, in no case, can work be accomplished by the transference of heat from cold to hot bodies. Hence the assumption that there can be a more efficient engine than a reversible one leads to a conclusion which we believe to be false.

Let us put the matter in another way. We have seen that our imaginary compound engine will do work. Let us use up this excess of energy by making it work another reversible engine backwards. On the whole, the system is now doing no work at all, and yet heat is being steadily transferred from a colder to a hotter body. This again is so against all experience that we are justified in saying that it is impossible.

Hence a reversible engine must be the best of all possible engines, and, therefore, all reversible engines are equally efficient.

Lord Kelvin, as far back as 1848, pointed out that we here have a basis for a scale of temperature which is independent of the properties of any form of matter. For since all reversible engines have the same efficiency, that efficiency must be independent of the nature of the working substance, and depend only on the temperatures of the hot and cold bodies. Now the efficiency is the ratio of the work done to the heat taken in, and the work done must, by the Principle of the Conservation of Energy, be equivalent to the difference between the quantity of heat taken in (Q_1) and that given out (Q_2). Thus the ratio $\dfrac{Q_1}{Q_2}$ must also depend only upon the temperatures of the two bodies. We can, therefore, define the ratio between the temperatures of those two bodies as the ratio $\dfrac{Q_1}{Q_2}$. It remains to be seen, however, if there is any close relation between the scale resulting from this definition and the property-of-matter scales in ordinary use.

One interesting and important consequence of this method of measuring temperature is that it leads naturally to the idea of a body containing no heat whatever, and the temperature of which, therefore, must be the lowest possible temperature.

For, suppose our reversible engine to deliver during a forward cycle no heat at all to Y. Here the ratio $\dfrac{\text{Heat taken from } X}{\text{Heat given to } Y}$ becomes infinite and thus,

$$\frac{\text{temperature of } X}{\text{temperature of } Y} = \infty .$$

This is only possible under the given conditions, when temperature of Y is zero. Thus not only the relative values of any two temperatures, but also the zero point, are defined for us on the absolute scale in a manner independent of all properties of matter.

In arriving at these conclusions no assumption has been made as to the size of the degrees on this scale. We may, therefore, choose any arbitrary value we please. If we prefer to call the difference between the temperature of steam under standard pressure and that of melting ice 100°, we can find how many of such degrees lie between freezing point and absolute zero. Suppose this number to be T then

$$\frac{\text{Absolute temperature of boiling point}}{\text{Absolute temperature of freezing point}} = \frac{T + 100}{T} = \frac{Q_1}{Q_2}$$

where Q_1 is the heat taken in, and Q_2 the heat given out by a reversible engine working between standard boiling and freezing points of water.

Now, if we assume as the working substance a perfect or ideal gas, viz. one which obeys Boyle's Law at all temperatures, we can show that the value of this ratio

$$\frac{Q_1}{Q_2} \text{ is (approximately) } \frac{373}{273}; \quad \therefore \frac{T + 100}{T} = \frac{373}{273};$$

thus $T = 273$ (more accurately 273·7).

Hence a body at $-273°·7$ C. is the coldest body possible.

Remember that this conclusion is based on the behaviour of an ideal gas, and it remains to be seen how far the gases we are acquainted with fulfil the above conditions.

I must not go into details of the comparisons that have been made; it is sufficient to say that the temperature-scale most closely in agreement with the absolute one, is that obtained by assuming that the pressure of a gas, such as hydrogen, is, at constant volume proportional to its temperature.

Scales thus constructed differ slightly according to the nature of the gas; hence the phrases *hydrogen scale*, *nitrogen scale*, etc. We may, however, consider it as established that when the gas is hydrogen the resulting scale differs so slightly from the absolute one at ordinary temperatures that it is sufficiently accurate for our purpose[1]. (For example, the absolute zero on this scale is at $- 273°\cdot13$.)

From what has been said, it appears that our definition of the C.G.S. primary unit must be as follows:

"If a reversible engine does 1 erg of work during a forward cycle when the temperature difference between the hot and the cold bodies is 1° of the absolute scale, then the primary thermal unit is the excess of the heat received over that ejected."

True, the unit thus indicated is a very small one: it would only raise the temperature of 1 gramme of water by about $\frac{1}{42}$ millionth of a degree. Fortunately, however, the usefulness of a thing is not proportional to its size (remember for example that the C.G.S. unit of potential difference is but $\frac{1}{10^8}$ of a volt).

It is now necessary to determine the value of our secondary watery-unit in terms of this primary one.

Our definition of this unit will be, " The number of primary units (or ergs) required to raise 1 grm. of water through 1° of the constant volume hydrogen thermometer at a fixed temperature on that thermometer."

Now, how is this number to be accurately ascertained? The first practical question is "in what way are we going to measure this 1° rise in temperature?" It is very easy

[1] See *Report of Electrical Standards Committee*, Appendix I., B. A., 1897.

to talk about finding it by means of a gas thermometer; but that is a most difficult instrument to use. I know of no physical apparatus which is so simple in theory, but so appallingly difficult in practice. A rough approximation is very easily obtained by means of it, but if the idea is to secure an accuracy of say $\frac{1}{1000}$th of a degree, the undertaking is a very serious one. Its direct employment in experimental work is rarely feasible, and the only course is to standardise a mercury, or other thermometer by comparisons with it, conducted under the most favourable circumstances, and then use that thermometer for the experimental work.

Even when a comparison with the hydrogen scale has been satisfactorily accomplished (and such cases are rare) sufficient allowance has often not been made for the erratic behaviour of all mercury-in-glass thermometers.

Our knowledge of the peculiarities of mercury thermometers has, in recent years, been greatly increased. The investigations of MM. Guillaume, Chappuis, and Pernet have not only indicated causes of error hitherto unsuspected, but have also given us methods of observation and correction by which it is possible to eliminate the effects of such errors. The determination of a single temperature by a mercury-in-glass thermometer is, however, a complicated and laborious one at the best. The following table is an example of such a determination[1] and it will be seen that some of the operations are of a kind that it would be difficult to carry out under ordinary circumstances.

[1] *Phil. Trans. Roy. Soc.*, Vol. 184, p. 429.

Temperature reading by thermometer P.

	millims.	Zero point millims.
Barometer (corrected)	754·1	754·1
Water pressure (in terms of mercury) ...	12·1	4·4
Total external pressure	766·2	758·5
Observed reading	14·025	− ·036
Calibration correction	− ·026	0
External pressure correction	− ·001	0
Internal „ „	+ ·026	+ ·008
Zero	+ ·028 ←——— − ·028	
Fundamental error correction	− ·006	

Sum corrections	+ ·021
Correction for stem temperature..	− ·002
Hence reading on mercury scale...	= 14·044
Correction to H scale − ·067 ; to N scale − ·059.	

Hence temperature is

Hydrogen scale.	Nitrogen scale.
13·977	13·985

The observations here recorded were made with a Tonnelot hard-glass thermometer. Had one of our ordinary English soft-glass thermometers been used, many of the corrections would have been larger and also more uncertain than those given in this Table.

If you reflect on all the preliminary labour which had to be accomplished before it was possible to apply these corrections, as for example in the standardisation, calibration, determination of the coefficients for changes in the internal and external pressures, the comparison with the gas scales, etc., I think you will admit that the popular belief in the simplicity and accuracy of the mercury-in-glass thermometer is based on ignorance rather than on knowledge.

The want of sufficient appreciation of these experimental difficulties has rendered useless much work that would otherwise have been of the greatest value.

Investigators seeking the value of the mechanical equivalent have taken elaborate precautions as regards the direct measurement of the work done, whereas they have paid too little attention to their thermometric standardisations and corrections. Even such masters as Joule and Rowland have erred in their thermometry to an extent which has seriously affected the accuracy of their conclusions.

In recent years a valuable weapon has been placed in our hands, viz. the platinum-resistance thermometer.

The change in the electrical resistance of a platinum wire, due to changes in temperature, is a quantity which can be determined with a high degree of accuracy, and although the resulting scale departs as largely from the hydrogen scale as does that of the mercury thermometer, the platinum thermometer has certain marked advantages. It is not erratic in its behaviour, its readings are more constant under varying conditions, and, by suitable arrangements, it renders evident much smaller temperature differences than could be determined by other methods. Hence, if the relation between the increase in electrical resistance of the platinum wire and the increase in temperature on the true temperature-scale has been accurately ascertained, the precision of temperature determinations obtained by the use of platinum thermometers exceeds, in my opinion, that obtainable by any other method; at all events at ordinary temperatures.

I have thought it necessary to dwell, at some length, upon this matter of the practical determination of temperature, as I propose, when discussing the values of the *mechanical equivalent* obtained by different observers, to attach great weight to the more recent determinations. Although, in some cases, the measurements of the work expended may be less satisfactory than in earlier experiments, the thermal measurements are of a higher order of accuracy.

In my first lecture I endeavoured to show you that it was a comparatively easy matter to completely convert any form of energy into heat.

To-day our enquiry into the meaning of the phrase ' a temperature scale' has led us to very important conclusions concerning the reverse process. We have seen that it is only possible to convert heat into mechanical work provided that we can obtain bodies at different temperatures, and further, that the fraction of the heat supply which can be thus converted is dependent on the temperature-difference. In other words, if we have a source of heat at temperature θ on the absolute scale and a condenser, or sink, at temperature t; then if we have a perfect engine at our disposal the maximum amount of heat we can utilise when any quantity of heat Q_1 is taken from the source, is $\dfrac{\theta - t}{\theta} \times Q_1$.

It is seldom that the difference of temperature between

the source and condenser of our steam-engines is as much as 120° C.; thus a perfect engine could only, under such circumstances, utilize less than ⅓ of the heat passed into it. I shall show you in my last lecture that, unless we can discover perfectly conducting materials as well as perfect non-conductors, it is certain that no real engine can approach to the efficiency of a reversible one. Thus, in practice, our engines transform into mechanical work but a very small fraction of the heat-energy which they receive.

In the phrase "*mechanical equivalent*," therefore, the meaning of the word "*equivalent*" is restricted to a numerical relation only.

If we say that 1400 ft.-lbs. are equivalent to 1 thermal unit, it does not follow that you would as soon have the one as the other. It is probable that you could make some profitable use of the 1400 ft.-lbs.; but it is only under exceptional circumstances that you could do anything whatever with your thermal unit.

I have here an apparatus which is probably the simplest form of heat-converting engine yet constructed. It more nearly resembles Carnot's engine in its action than any other engine I know; but, at the same time, the differences are vital, and I am afraid the efficiency of this engine is but a small fraction of that of a reversible one. At the same time, we have here no valves and the contents are in the same condition after, as before, a cycle; only, unfortunately, those contents have been constantly in contact with conducting bodies at different temperatures and, therefore, it would be of no use to compare the quantities of heat received and ejected, even if we were able to do so.

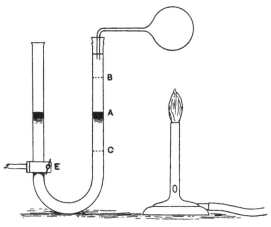

Fig. 4.

The apparatus consists of a large **U** tube partially filled with mercury. A perforated cork is inserted at one end of this tube and the neck of a glass bulb passes through the cork. Thus, when the surface of the mercury at A descends, some of the air in the bulb passes into the upper part of the **U** tube, and *vice versâ*. If the flame of a Bunsen burner be now placed beneath the bulb, the column of mercury begins to oscillate, the amplitude of the oscillations being considerable[1], and although the

---

[1] In the apparatus shown at Leeds the amplitude of the oscillations was about 8 cm.

centre of gravity of the mercury is not permanently raised, it is obvious that work must be done in maintaining this oscillation against the frictional resistance[1].

As the mercury passes down from B to C the space thus vacated is filled by hot air from the bulb. This air is cooled both by the work it has done and by the contact with the cold surfaces in the tube. As the

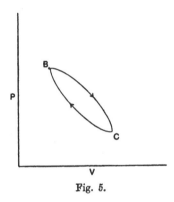

Fig. 5.

mercury rises from C to B this air, whose temperature will now have been considerably reduced, is passing back into the bulb. Hence we see that the average temperature and, therefore, the average pressure of the air is greater during the descent than during the ascent of the mercury in the closed limb. Hence in a complete cycle

[1] An oscillation of this kind was observed some years ago by Mr A. Vernon Harcourt when passing electric sparks through a mixture of hydrogen and nitrogen confined above a column of mercury. At first he attributed the effect to an explosive action, but further experiments proved that the movement was maintained when other gases were used and the heat applied externally.

more work is done on the mercury as it is forced from
B to *C* than it does upon the gas during its return, and
therefore the movement can be kept up in spite of
friction, etc.; the indicator diagram being probably somewhat of the type shown in Fig. 5.

The oscillation can only continue so long as the
temperature of the bulb exceeds that of the tube. The
movement of the contained air during each stroke tends
to diminish this difference and, therefore, the motion
would soon cease, but for the constant loss of heat by
conduction and radiation from the tube, and the gain of
heat by the bulb.

The action of hot-air engines is of the same nature,
although it is obscured by the many moving pieces.
Considering the small difference in temperature that
must exist after they have been running for a considerable time, the power developed by such engines
is surprisingly great.

I am afraid that to-day's lecture may, to some of those
present, have appeared a needless repetition of truths
already known and appreciated. I have, nevertheless,
ventured to inflict it upon you, for I believe that the
difficulties of temperature measurements are, as a rule,
greatly underestimated. Whatever views we may hold
regarding those difficulties, it is, however, impossible to
deny the importance of this subject when we reflect
on the number of physical constants whose accurate
determination is dependent upon the measurement of
temperature.

NOTE. At the close of this lecture two rough experimental determinations of the '*mechanical equivalent*' were performed as an introduction to the methods of the observers whose work I proposed to discuss during the following lecture. I have in Appendix II. given an account of those experiments in the hope that it may be found useful by teachers, although neither the apparatus nor the methods present any specially novel features.

LECTURE III.

Table of values of *J* obtained by different observers. Direct and Indirect Methods of Measurement. Principles which should guide us when making a selection. Brief descriptions of the Methods of Joule, Hirn, Rowland, Reynolds and Moorby, Griffiths, Schuster and Gannon, Callendar and Barnes. Table of Results.

IN my first Lecture I called your attention to the transformation of various forms of energy into heat, a process involving what is commonly called *degradation* of energy. In the second Lecture our enquiry as to the nature of a temperature scale led to the consideration of the converse operation, a transformation involving what we may term *elevation* of energy.

We found that the process of degradation could be accomplished with comparative ease; whereas, at all events under the conditions holding on our earth, the process of elevation is one which presents peculiar difficulties.

Now, unfortunately, or possibly fortunately (for the sight of brave men struggling with adversity is grateful to the Gods), it is the transmutation from the lower to the more exalted forms of energy which almost entirely engages the attention of our engineers.

The physicist, however, is not hampered by the conditions which bind the engineer; hence, in the determinations of the mechanical equivalent, he has, in general, trodden the downward path.

The experimental investigations may be placed under two headings, Firstly those in which the thermal effect is directly produced by the expenditure of mechanical work, and secondly where it is due to the expenditure of energy other than mechanical. In the latter case it is necessary that the numerical value of this energy should already be known in dynamical units; as, for example, where the work is done by means of an electric current. The first may be termed the direct, and the second the indirect method.

The direct method naturally carries greater weight when our one object is the determination of the numerical constant giving the relation between work done and equivalent heat; but the indirect methods are also of high importance, for they enable us to test the accuracy of various physical standards, such as electrical resistance etc., and also to express all forms of energy in terms of a common denominator.

Tables I. and II. are (with the exception of the results of experiments completed since 1893) taken from Preston's *Theory of Heat,* and I believe they contain all the determinations that have any pretensions to accuracy.

Unfortunately the values in this table are all given in kilogramme-metres, and therefore no exact comparison between them is possible unless the value of g is known at each place where the observations were conducted; for

TABLE I.   (DIRECT METHODS.)

Date	Observer	Method	Result
1843	Joule	Friction of water in tubes ...	424·6 km.
,,	,,	Electromagnetic currents ...	460
,,	,,	Decrease of heat produced by a pile when the current does work	442·2
1845	,,	Compression of air	443·8
,,	,,	Expansion of air	437·8
,,	,,	Friction of water in a calorimeter	488·3
1847	,,	,, ,, ,, ,,	428·9
1850	,,	,, ,, ,, ,,	423·9
,,	,,	Friction of mercury in a calorimeter	424·7
,,	,,	Friction of iron plates in a calorimeter	425·2
1857	Favre	Decrease of heat produced by a pile doing work	424·464
,,	Hirn	Friction of metals	371·6
1858	,,	,, ,, ...	400·450
,,	Favre	Friction of metals in mercury calorimeter	413·2
,,	Hirn	Boring of metals	425
1860–61	,,	Water in friction balance ...	432
,,	,,	Escape of liquids under high pressure	432, 433
,,	,,	Hammering lead	425
,,	,,	Friction of water in two cylinders	432
,,	,,	Expansion of air	440
,,	,,	Steam-engines	420·432
1865	Edlund	Expansion and contraction of metals ... ... 428·3, 443·6	
1870	Violle	Heating of a disc between the poles of a magnet	435
1875	Puluj	Friction of metals ... ... 425·2, 426·6	
1878	Joule	Friction of water	423·9

TABLE I. (*continued*).

Date	Observer	Method	Result
1878	Rowland	Friction of water between 5° and 36°	... 429·8, 425·8
1891	D'Arsonval	Heating of a cylinder in a magnetic field	421·427
1892	Miculescu	Friction of water	426·84
1897	Reynolds and Moorby	,, ,, Mean capacity 0° to 100°	426·27

TABLE II. (INDIRECT METHODS.)

Date	Observer	Method	Result
1842	Mayer	By the relation of $J = \dfrac{p_0 v_0 a}{c_p - c_v}$ for gases	365 km
1857	Quintus Icilius	Heat developed in a wire of known resistance	399·7
,,	Weber	Heat due to electric currents ...	432·1
,,	Favre ⎱ Silberman ⎰	Heat developed by zinc on sulphate of copper	432·1
,,	Bosscha	Measure of E.M.F. of a Daniell's cell	432·1
1859	Joule	Heat developed in a Daniell's cell	419·5
,,	Bosscha	E.M.F. of a Daniell's cell ...	419·5
,,	Lenz-Weber	Heat developed in wire of known resistance	... 396·4, 478·2
1867	Joule	,, ,, ,, ,,	429·5
1878	Weber	,, ,, ,, ,,	428·15
1888	Perot	By the relation $L = \tau (v_2 - v_1) \dfrac{dp}{dt}$	424·63
1889	Dieterici	Heat of electric currents ...	432·5
1893	Griffiths	,, ,, ,, ...	427·45
1894	Schuster and Gannon	Electric current, E. & C. being known	427·19
1899	Callendar and Barnes	,, ,, ,, ,,	426·52

I find by one or two reductions that they have not been corrected to a common latitude. If the English ones are multiplied by $981\cdot24 \times 100$, and the French by $980\cdot94 \times 100$, the result will give the value in ergs with a sufficiently close approximation for the purposes of a rough comparison, for it must be remembered that the temperature scales differ considerably, as also the mean temperature of the temperature range.

In these tables the values vary from 365 to 488 kms.; how then are we to decide which to select?

There are some would-be physicists who apparently believe that if you obtain a sufficient number of observations and find the mean (especially if you apply the method of least squares) you will probably be about right. I do not see, however, that there would be any advantage in following this course where we have to deal with such divergent results as these. It should ever be remembered that a few good experiments are probably better than the mean of one hundred faulty ones, for it is quite possible that the errors of all the faulty ones are in the same direction.

To assist us in making a selection, I propose to ask the following questions:

(1)   Are the temperature determinations sufficiently accurate?

The answer will lead to wholesale rejection, especially in the earlier experiments.

As I have previously indicated, the difficulty and importance of temperature measurements were not sufficiently appreciated until within very recent times, and, unfortunately, an error in thermometry is, as a rule, a fatal one; for each thermometer has its own peculiarities and special causes of error; thus, no later increase in knowledge enables us to correct results unless the actual thermometers have been preserved and the conditions, under which they were used, fully recorded.

Fortunately in two of the most important cases (viz. Joule's and Rowland's) the thermometers actually used have been preserved.

In the former, however, our information is not complete for we are not sufficiently acquainted with the exact conditions under which their readings were observed by Joule. In the latter, a restandardisation has been accomplished under Rowland's own direction, and thus the corrections can here be applied with far greater certainty. In both cases, as I shall show you later, the results as originally published have, in consequence, undergone considerable modification.

(2) Has the author given us sufficient data to enable us to judge the probable accuracy of all the various measurements involved by his method of experiment?

In this respect, also, the earlier determinations are at a disadvantage, as compared with the later ones, for the importance of full information concerning the details of physical measurement has only been generally recognised in recent times.

(3) Are we certain that the energy of the bodies under observation has undergone no modification during the experiment in consequence of molecular changes?

If we could accurately determine both the kinetic

energy expended in hammering a nail and, also, the heat developed, it is not certain that the resulting value of the constant would be correct; for the condition as regards density, strain, &c. of the nail (and possibly of the hammer-head) might have undergone alteration and, in consequence of new molecular conditions, have gained, or lost, in energy.

Now, we know that no permanent shearing strain can exist in a fluid, and if the external pressure is unaltered, the density will have undergone no change except that due to change of temperature. Hence, conclusions drawn from observation of work expended in heating a liquid are, *cæteris paribus*, of leading importance.

I will not trouble you with further details of the considerations which have led to the selection of the experiments which I am about to discuss. Suffice it to say that a careful study of the writings of most of those authors who are mentioned in these tables has led to the selection from table I. of the work of Joule, Rowland, and Reynolds and Moorby, and from table II. the determinations of Griffiths, Schuster and Gannon, and Callendar and Barnes[1].

I believe that the results obtained by Rowland, after the revision of his thermometry, should be considered as of leading importance in the estimation of the

---

[1] Those who may wish to obtain information concerning the work of the other observers mentioned in Tables I. and II. should consult the summary given by Prof. Ames in the *Rapports présentés au Congrès International de Physique*, Paris, 1900, Tome I.

In my criticisms on the works of the above authors I have ventured to quote largely from this excellent Report.

numerical value of the constant; while the indirect methods of Griffiths, and above all of Callendar and Barnes enable us to trace the changes in the capacity for heat of water and thus render it possible to make a comparison of the values obtained by the different observers in terms of the constant primary unit. I propose, therefore, to consider in some detail the determinations I have mentioned and also, for special reasons, the work of Hirn.

JOULE[1]. The form of apparatus first used by Joule (in which the descent of a weight caused the rotation of a paddle) is the one ordinarily given in the text-books and is, I am sure, familiar to all.

Now, text-books are curious things. If a description of a piece of apparatus, or the value of any natural constant once appears in *any* text-book, it is apparently a law that it should continue to appear in *all* text-books for the remainder of time. Thus the apparatus used by Joule in 1845-7 is almost invariably presented to us, while the improved type used by him for his later, and more accurate, work is generally ignored.

The method finally adopted by Joule consisted in stirring water by means of a paddle which was rapidly turned by hand-wheels, shown at *d* and *e* (Fig. 6); the vessel was suspended by a vertical shaft *b*, which also carried a large fly-wheel *f*. The mass of the water and the water-equivalent of the calorimeter were carefully determined, the rise in temperature was noted on a mercury-in-glass

[1] *Scientific Papers*, vol. I. pp. 632—657.

thermometer, and the work consumed in heating the water was measured by a dynamometer, which consisted of an arrangement for balancing the moment acting on the suspended calorimeter (owing to the revolution of the paddle) by a moment produced by the tension of the

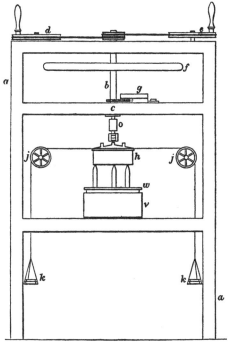

Fig. 6.

cords fastened tangentially to the calorimeter. The cords passed over pulleys and supported weights k. If this moment is constant and is called M, and if the number of revolutions per second of the paddle is N, the work done per second is $2\pi MN$.

In order to reduce the metallic friction as far as possible, the base of the calorimeter rested on a hydraulic supporter, which consisted of two concentric vessels v and w, the space between them being filled with water. The three uprights attached to w pressed against the base of the calorimeter and reduced the pressure on the bearing at o nearly to zero.

Joule's calorimeter had a water equivalent of 313·7 grammes of water at 15·5° C.; the mass of water used in an experiment was about 5124 grammes, each experiment lasted 41 minutes and the observed rise in temperature was about 2·8° C. The mean of his results gave 772·65 foot-pounds at Manchester, as the quantity of work required to raise the temperature of one pound of water 1 degree F. on his mercury-in-glass scale at 61·69° F. Changing to the centigrade scale and to the C.G.S. system, Joule's result may be stated as follows:—the quantity of work required to raise the temperature of 1 gramme of water 1 degree centigrade on his mercury-in-glass thermometer at 16·5° is $4·167 \times 10^7$ ergs.

In 1895 Professor Schuster[1] compared Joule's thermometer with a Tonnelot thermometer which had been standardised in terms of the nitrogen thermometer of the Bureau International at Sèvres; and in this way was able to recalculate Joule's value for the mechanical equivalent. Rowland also, when reviewing this experiment of Joule's, called attention to certain errors in the

[1] *Phil. Mag.* xxxix. pp. 477—506. 1895.

determination of the water-equivalent of the calorimeter, whose value had, in consequence, been underestimated by nearly 1 part in 1000. The corrected result is given by Schuster as follows: the specific heat of water at 16·5° C. Paris nitrogen scale is $4·173 \times 10^7$ ergs.

This value differs by 1 part in 400 from determinations made by Rowland and others. In fact the thirty-five observations from which Joule calculated his final value give results which differ by more than 1 per cent. from each other, and his thermometer did not permit the most accurate reading of temperature. The cause of these discrepancies is to be found in three conditions of the experiment: irregularity of the stirring, which was done by hand, incomplete correction for loss of heat by radiation, and insufficient knowledge of the variations in the readings of mercury-in-glass thermometers caused by variations in the conditions under which they are used and standardised.

Let us now on account of its intrinsic interest consider briefly the work of HIRN[1]. I may say at once that it is impossible to attach any importance to his numerical results. His thermometry was far too vague and his energy-measurements were also not of the highest accuracy. Nevertheless, his work is important; because, not only was his method of degrading energy entirely different from that of Joule but he also endeavoured to obtain the value of the equivalent by the reverse process, that is by measuring the disappearance of heat from a

[1] *Théorie Mécanique de la Chaleur.*

system performing work; a feat, I believe, only seriously attempted by Hirn. The method he adopted in the degradation process was as follows:

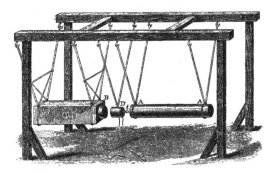

Fig. 7.

A cylinder of iron *AA* (Fig. 7), weighing 350 kilos, was suspended by two pairs of cords which compelled it to move in a vertical plane with its axis always horizontal. This cylinder was used as the hammer or instrument of percussion. The anvil *MB* was a large prismatic mass of stone weighing 941 kilos, and suspended in the same way as the hammer. The mass of lead *D* to be operated on was suspended between the two, and the face *B* of the anvil adjacent to the lead was cased with iron to receive the blow. In making an experiment the hammer was drawn back by a tackle, and the height to which it was raised was accurately measured. It was then let fall upon the lead, and the recoil of the anvil was registered by a sliding indicator which was pushed back and then remained *in situ*. An observer also noted the advance or recoil of the hammer after the blow, and from these data the work spent in percussion could easily be calculated, as the difference between the kinetic energy of the system before and after impact was known. Before the blow was delivered, the temperature of the lead was taken by inserting a thermometer t into a cylindrical cavity made in the mass, and immediately after the blow the mass of lead was removed and hung up by two strings provided for the purpose, so that the axis of the cavity was vertical. This cavity was immediately filled with ice-cold water, which was stirred and the rise of temperature noted. The value 425 kgs. was obtained by Hirn in this manner, which is remarkably good considering the nature of the experiment.

His more interesting enquiry, however, was an attempt to determine the difference between the heat conveyed from the boiler to the condenser of an ordinary steam-engine when running first with a heavy, and afterwards with a light load.

After making allowance for the heat lost by radiation, conduction, etc. he proved that the difference was much greater when the engine was performing heavy external work than when it was running with no load.

" Hirn also pushed the experimental enquiry further, and actually deduced a fair estimate of the dynamical equivalent of heat from the observations of the work done by the engine, and the quantity of heat used up in performing it. The work performed in any time can be deduced from the area of the Watt's indicator diagram, and the number of strokes of the piston. To determine the quantity of heat converted into work, the weight of water that passes from the boiler to the condenser must be estimated. Knowing the latent heat of vaporisation at the temperature of the boiler, the quantity of heat Q drawn from the boiler in any time becomes known. But this quantity is not all converted into work. Part of it q is carried into the condenser, and a part R is lost by radiation in the transit. Hence the quantity of heat converted into work is $Q - q - R$, and the value of J is found from the equation

$$W = J(Q - q - R).$$

" By this means Hirn obtained the numbers 413 and 420·4 (gramme-metres), which, considering the difficulty of the investigation, must be regarded as exceedingly good approximations[1]."

These results are extremely interesting; but, as I said before, I do not propose to include the actual numerical values in our final table.

I now pass to what is undoubtedly the most important determination of all, namely, ROWLAND'S[2].

In the years 1878 and 1879 Professor Rowland of the Johns Hopkins University, Baltimore, performed a series of experiments for the determination of the specific heat of water at different temperatures. His method was in principle the same as that of Joule, although devised independently.

By means of a petroleum engine, a specially designed paddle-wheel was turned at a rapid rate (200 to 250

[1] Preston's *Theory of Heat*, p. 47.
[2] *Proc. American Academy*, No. 1880—81.

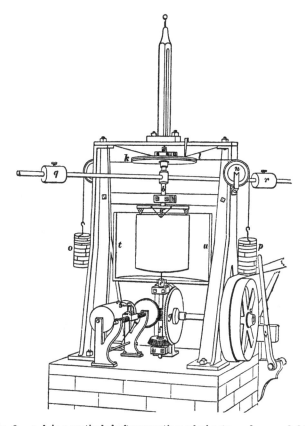

Fig. 8. *a, b* is a vertical shaft supporting calorimeter and suspended by a torsion wire.

Axis of paddle passed through base of calorimeter and was connected with shaft *e, f*, which was kept in uniform rotation by the driving engine.

o and *p*, weights attached to silk tapes passing round wheel *k, l*, the couple acting on calorimeter being thus measured (corrections being applied for the torsion of the suspending wire).   The moment of inertia could be varied by means of the weights *q* and *r*.   A water-jacket, *u, t*, surrounded the calorimeter and was used for the estimation of the radiation.

revolutions per minute) in a calorimeter which was suspended by a wire and which was prevented from turning, when the paddle was revolving, by means of a moment applied by weights. The water equivalent of the calorimeter was 347 grammes, and the rise in temperature was about 0·6° C. per minute, and was observed on different mercury-in-glass thermometers. These thermometers had all been compared with a constant volume air-thermometer, and the readings were all finally reduced to the "absolute scale"; the results of Thomson and Joule's experiments on the expansion of air through a porous plug being used to make the necessary corrections to the air-thermometer temperatures. While the temperature was rising through any range, e.g. from 8° to 30° C., readings were made on all the instruments for each degree or half degree, and in this way any one experiment gave a great number of determinations of the capacity for heat. The work was calculated for 10° intervals, and one-tenth of that required to raise the temperature of one gramme of water from $(t-5)°$ to $(t+5)°$ was called the specific heat at $t°$ C. All corrections for losses due to radiation, variations in speed of paddle, &c., were carefully considered and made.

Rowland's method of using his mercury thermometers was different from that which has been universally adopted within the past few years owing to the efforts of Prof. Pernet of Zürich, and of MM. Chappuis and Guillaume of the Bureau International, and the scale of his air-thermometer in terms of other thermometers used by later observers was not known. In 1897, therefore, a series of comparisons was undertaken at the Johns Hopkins University between Rowland's thermometers, three Tonnelot mercury thermometers standardised at the Bureau International, and a Callendar-Griffiths platinum thermo-

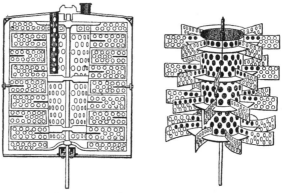

Fig. 9. Section of Rowland's Calorimeter and sketch of Paddle.

meter. The result has been a recalculation of Rowland's figures. In the following table the values for the specific heat are given as Rowland first published them, and as recalculated by Day[1], and by Waidner and Mallory[2].

Professor Pernet has also endeavoured to recalculate Rowland's values from a careful study of Baudin thermometers of the same glass and construction as those of Rowland. His figures are almost exactly 1 part in 400 less than those determined by Day and by Waidner and Mallory, and in such a case as this, one must have the greater confidence in the direct comparisons.

[1] *Phil. Mag.* XLVI. 1—29, 1898.
[2] *Physical Review*, VIII. 193—236, 1899.

TABLE III.

Tempera-ture	Rowland's original values, absolute scale	Recalculated by Day, Paris hydrogen scale	Same reduced to Paris nitrogen scale	Recalculated by Waidner and Mallory, Paris nitrogen scale
5° C.	4.212×10^7 ergs			
10	4.200	4.196×10^7	4.194×10^7	4.195×10^7
15	4.189	4.188	4.186	4.187
20	4.179	4.181	4.180	4.181
25	4.173	4.176	4.176	4.176
30	4.171	4.174	4.174	4.175
35	4.173	4.175	4.175	4.177

Rowland took pains to vary all the conditions of his experiments as much as possible, running his engines at different speeds, using different thermometers, carrying his observations over different ranges and making in all thirty series of observations. Therefore great weight must be given to his determinations. The only criticisms that can be made are that the range of 10 degrees is too large if the specific heat at the mean temperature is desired, and that the radiation correction is uncertain at and above 30°. As Rowland himself says: "The error due to radiation is nearly neutralized, at least between 0° and 30°, by using the jacket at different temperatures. There may be an error of a small amount at that point (30°) in the direction of making the mechanical equivalent too great, and the specific heat may keep on decreasing to even 40°."

Professor Ames's criticism on this work is as follows:

"Rowland estimates his possible error at less than two parts in 1000; and, now that his thermometric readings have been recalculated, the possible error is probably reduced to less than one part in 1000, unless there was a constant or systematic error, which is most improbable. Rowland's method of making thermometric readings is one which is liable to serious error, and it is possible that in the recalculations made by Day, and by Waidner and Mallory, the thermometers were not used in identically the same manner as they were originally. There is no obvious reason, however, for believing, as Pernet does, that there is a systematic error in Rowland's research."

The experiments of REYNOLDS and MOORBY[1] are also examples of physical work of the highest order. In 1897 they published an account of their determinations of the mean specific heat of water between 0° and 100° C. The apparatus used was of such a nature that I do not think it possible to convey, in a brief description, any clear idea of the machinery and its connections; therefore, I will only call your attention to the manner in which the work was controlled and estimated. The general idea is that of a hydraulic brake attached to the shaft of a triple expansion 100 horse-power engine making 300 revolutions per minute.

The water enters the brake at, or near, 0° C. and runs through it at such a rate that it issues at, or near, 100° C., the work expended on the water being estimated by means of a dynamometer consisting of a lever and weights fastened to the brake. The whole of the work done being

[1] *Phil. Trans.* A. 1897.

absorbed by the agitation of the water in the brake, the moment of resistance of the brake at any speed is a definite function of the quantity of water in it. Except for this moment the unloaded brake is balanced on the shaft, the load being suspended on the brake lever at a distance of 4 ft. from the axis of the shaft. If the moment of resistance of the brake exceeds the moment

Fig. 10. This figure shows the hydraulic brake, lagged with cotton-wool covered by flannel, the brake lever projecting from it towards the left. The load carried by the lever can be seen at the bottom left-hand corner.

of this load, the lever rises and *vice versâ*. By making this lever actuate the valve which regulates the discharge from the brake, the quantity of water is continually regulated to that which is just required to support the

load with the lever horizontal, and thus a constant moment
of resistance is maintained whatever the speed of the
engines.

In order to eliminate as many errors as possible, three
" heavy trials " were made in succession, followed by three
" light trials," each trial lasting 60 minutes, and the
difference in the two cases, " heavy " and " light," of the
' mean work per trial ' and the difference of the ' mean
heat per trial ' were taken as equivalent.

In a " heavy trial," the dynamometer was adjusted to
a moment of 1200 foot-pounds, and the quantity of water
run through in the 60 minutes was about 960 pounds ;
in a " light trial " the moment was generally 600 foot-
pounds, and the quantity of water run through in the
60 minutes was about 475 pounds, although six trials
were made with the moment at 400 foot-pounds. As it
was only necessary to determine temperatures in the
neighbourhood of 0° and 100° C., the results are almost
independent of the *nature* of the temperature scale, as all
temperature scales must be in agreement at the two
standardising points, while the temperature range was so
great that an error in actual elevation at either' end of it
would have but a small effect. Again, the large scale on
which the experiments were conducted would tend to
diminish the effect of inaccuracies in the measurements of
the thermal loss by radiation, etc. The most minute
attention was paid to all possible causes of inaccuracy, and
there is no apparent constant source of error in the final
results.

When their value is expressed in ergs, it becomes

4·1833 × 10⁷; that is, the mean capacity for heat of unit mass of water between 0° and 100° C. is $4·1833 \times 10^7$.

Unfortunately the work of Reynolds and Moorby does not afford us much assistance in our efforts to determine the actual value of the heat equivalent. Assuming the validity of their conclusions, we know how many ergs (or primary units) are required to raise the temperature of 1 gramme of water from 0° to 100°; but we are unable to compare their results with the values obtained by Joule and Rowland, unless we know the relation of the mean thermal unit over the range 0° to 100° to the thermal unit at the temperatures covered by the experiments of those observers. It is quite certain that the number of primary units required to raise 1 gramme of water through 1° at different temperatures is not the same; hence it is impossible in the present state of our knowledge to ascertain if Reynolds and Moorby's results are coincident with those obtained by other investigators.

On the other hand, if we assume the validity of the result obtained by Reynolds and Moorby, and compare it with the numbers given by Rowland, we can find the value of the mean thermal unit in terms of a thermal unit at some definite temperature;—a piece of information of great value when we remember that no use can be made of the Bunsen's calorimeter methods in the absence of such knowledge.

I now pass to the consideration of indirect methods, many of which were first employed by Joule. At various times he estimated the heat developed by electro-magnetic currents, the decrease of heat in a voltaic cell when the current does work, compression and expansion of air, &c.

I have already called your attention to the great variety of experiments performed by Joule, and it would be difficult to overestimate the importance of this variety as evidence that all forms of energy can be expressed as heat. For the purpose of our present enquiry, however, viz. the exact determination of the equivalent, his indirect determinations are of little use.

Next, in chronological order, I come to my own experiments during the years 1888 to 1893[1].

It is difficult, of course, for a parent to speak dispassionately concerning one of his own children and, therefore, I propose to quote the remarks of Prof. Ames, of Baltimore, in his Report addressed to the International Congress of Paris in 1900. I should, however, like to make two preliminary statements.

Firstly, my chief object in embarking on this investigation was to ascertain the validity of our system of electrical units.

If all the measurements were correct by means of which the practical values of the volt, the ohm, and the ampère had been determined, then we know that if the ends of a conductor, whose resistance is 1 ohm, are maintained at a potential difference of 1 volt, the work done by the current per second must be 10^7 ergs, and if this work is wholly expended as heat, then if J be

[1] *Phil. Trans. Roy. Soc.* A. 1893; *Proc. Roy. Soc.* LV. 1894; *Phil. Mag.* XL. 1895.

the ratio of the secondary to the primary heat unit, the heat liberated should be $\dfrac{10^7}{J}$ secondary heat units per second, where J is the ratio of the secondary to the primary heat unit.

Of course if we start by assuming the validity of our electrical measurements, we can thus obtain the value of J; but, I repeat, my object was to proceed in the opposite direction.

Secondly, I was anxious to throw some light on the changes in the capacity for heat of water, for, at that time the only information of importance that we possessed in this matter was that due to Regnault and Rowland, and, while results deduced from the work of the former indicated a steady increase from 0° upwards, Rowland had found a decrease over the range 5° to 30° C.

I may add that of the three or four years I expended on this work the greater portion was employed in the standardisation of the thermometers.

Prof. Ames writes as follows:

" Mr E. H. GRIFFITHS, of Cambridge, England, devised a method for the determination of the specific heat of water by the use of the heating effect of an electric current, which is, to a large extent, free from the errors connected with the previous methods in which electric currents were used.

" If a coil of wire carrying a current is immersed in water any one of the three following methods may be used to determine the energy spent in raising the temperature of the water:

(1) measure E, C and t.
(2) measure C, R and t.
(3) measure E, R and t.

" Griffiths used the third method, although for many reasons it is the most difficult. The obvious difficulty lies in the measurement of R; because, unless measured actually during the progress of the heating experiment, it is necessary to know the temperature of the wire and the temperature coefficient of its resistance; and its temperature is *not* that of the surrounding water. Griffiths thought to obviate this difficulty by making a series of subsidiary experiments which were designed to give the difference in temperature between the water and the wire when the former was at a known temperature and an E.M.F. of known strength was applied to the latter. The resistance of the wire was then measured at a known temperature and its temperature coefficient was also measured; therefore, when in the course of a heating experiment the temperature of the water was read, the resistance of the wire could be calculated. Griffiths found also that most rapid and thorough stirring of the water was necessary in order to secure consistent or satisfactory results. He designed a most efficient stirrer which made about 2000 revolutions per minute, the rise in temperature produced by the stirrer alone being in some cases equivalent to 10 per cent. of the whole work spent in raising the temperature. The necessary correction, owing to this, was ascertained by a series of preliminary experiments.

" Griffiths' apparatus consisted of a platinum wire (diameter 0·004 in. (0·010 cm.), length 13 in. (33 cm.), resistance about 9 ohms) coiled inside a cylindrical calorimeter, 8 cm. in height and 8 cm. diameter, whose water equivalent was 85. This wire was heated by means of a current from storage cells. The terminals of the wire

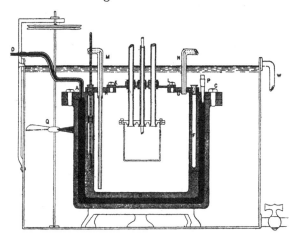

Fig. 11. Section of constant temperature chamber in which the calorimeter was suspended by glass tubes. *A B C* is a large steel vessel with double walls, the annular space (printed black in figure) being filled with mercury which is connected with a gas regulator by the tube *D*. The steel vessel stood in a large tank filled with water, which was rapidly stirred by the paddle *Q*. A small stream of water flowed continuously into the tank, the excess passing away at *W*. The temperature of the incoming water was controlled by the regulator which was governed by the mass of mercury (exceeding 70 lbs.) within the walls *ABC*. A very constant temperature could thus be maintained within the steel vessel. The space between the calorimeter and the steel walls was thoroughly dried and the pressure reduced to less than 1 mm.

were maintained at a constant difference of potential by balancing against sets of Clark cells; and, while the temperature of the water contained in the calorimeter was raised from 14° to 25° C., the time varying from forty

to eighty minutes, observations of the temperature and
time were made every degree. The E.M.F. used varied
from that of three to six Clark cells. Experiments
were made using different quantities of water; and by
taking *differences* in the energy and the heat produced
in the different sets, many errors were eliminated,
and the water-equivalent of the calorimeter disappeared
from the equation. In the end, therefore, the method
depended on the introduction of 120 grams of water into
the calorimeter, this being the difference between the
quantities used in two trials.

"Griffiths measured his E.M.F. in terms of the Cavendish
standard Clark cell; his resistance in terms of the "B. A.
ohm of 1892," which is the "International Ohm" as
defined in 1893; his time by a "rated" chronometer; and
his temperature by a Hicks mercury thermometer which
had been compared with a Callendar-Griffiths platinum
thermometer and also with a Tonnelot thermometer
standardised at the Bureau International. In accordance
with the work of Glazebrook and Skinner he assumed the
E.M.F. of the Clark cell at 15° C. to be 1·4344 volts, and
its temperature coefficient to be $1 + 0·00077 (15 - t)$.

"Later, Schuster called attention to an error of one
part in four thousand due to the capacity for heat of the
displaced air[1]; but this was neutralized by the fact that
there was a slight probable error discovered in the
estimation of the E.M.F. of the Clark cells used by
Griffiths[2], reducing the value to 1·4342 volts at 15° C.

[1] *Phil. Trans. Roy. Soc.* A. 1895.
[2] *Phil. Mag.* XL. 1895.

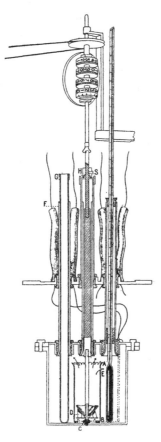

Fig. 12. Section of calorimeter showing the stirring arrangements at *D* by which the water was drawn through the bottom of the cylinder *AB*, and thrown against the roof of the calorimeter. The platinum coil is not shown in this section, as it was wound in a horizontal circle placed near the base of the vessel, the rack on which it was wound being supported by the rod indicated by the dotted lines to the right of the thermometer bulb. At the top of the stirrer-shaft is shown the counter which recorded the number of revolutions.

"Hence Griffiths' final values are :

15° C. nitrogen scale, $4\cdot198 \times 10^7$ ergs.

20°   ,,   ,, $4\cdot192 \times 10^7$,,

25°   ,,   ,, $4\cdot187 \times 10^7$,,

"In criticism of the method, it may be said that using as small quantities of water as Griffiths did, always practically under the same external conditions, there is more opportunity than should be for systematic errors and for errors due to radiation corrections. In this connection reference must be made to criticisms by Schuster[1] and to the reply by Griffiths[2]."

It will be seen from this summary by Professor Ames that the results of my work lead to the conclusion that (assuming the value of J obtained from Rowland's revised experiments) there is at all events no error exceeding 1 in 1000 in the value of our electrical units; but that there is an indication of a possible error of some such magnitude in the electro-chemical equivalent of silver, an indication which (as we shall see later) is strengthened by the work of Schuster and Gannon, Kahle, Patterson and Guthe.

A comparison of the curves resulting from Rowland's work and my own proved that we could not both be correct in our thermometry, and this of course excited suspicion as to any conclusions regarding the value of the electrical units. The revision of Rowland's thermometry was, I believe, partly due to this discrepancy, and the double restandardisations by different methods were, as

---

[1] *Phil. Trans.* CLXXXVI. A. 1895.

[2] *Phil. Mag.* 1895. pp. 431—454.

we have seen, Table III. *supra*, in agreement with each other, while the resulting corrections caused Rowland's curve of the changes in capacity for heat to be almost parallel with my own over the range of my experiments. This parallelism did not mean that the value of J, obtained by the assumption of the validity of the electrical units, was coincident with Rowland's, but that our remaining differences were probably due to the nature of the calorimetric determinations, or to some hitherto undiscovered error in our system of electrical measurements.

The results obtained by Professor SCHUSTER and Mr GANNON were published in 1895[1].

They also measured the heat developed by an electric current when overcoming resistance; but in this case, the work was estimated by observation of E and C, the latter by the use of a silver voltameter. (See Fig. 13.)

The rise in temperature was determined by a Baudin mercury thermometer, which was compared directly with a Tonnelot thermometer standardised at the Bureau International. The calorimeter had a water equivalent of 27 and the mass of water used was about 1514 grammes. The heated wire was of platinoid, 760 cm. long and of about 31 ohms resistance. The E.M.F. was produced by storage cells, and was constantly balanced against twenty Clark cells. The resulting current was in the neighbourhood of 0·9 ampere, and passed in series through a silver voltameter consisting of a silver plate and a platinum bowl

[1] *Phil. Trans. Roy. Soc.* CLXXXVI. A. 1895.

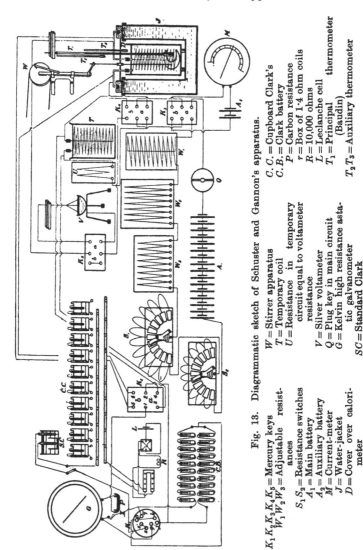

Fig. 13. Diagrammatic sketch of Schuster and Gannon's apparatus.

$K_1 K_2 K_3 K_4 K_5$ = Mercury keys
$W_1 W_2 W_3$ = Adjustable resistances
$S_1 S_2$ = Resistance switches
A_1 = Main battery
A_2 = Auxiliary battery
M = Current-meter
J = Water-jacket
D = Cover over calorimeter

W = Stirrer apparatus
T = Temporary coil
U = Resistance in temporary circuit equal to voltameter resistance
V = Silver voltameter
Q = Plug key in main circuit
G = Kelvin high resistance astatic galvanometer
SC = Standard Clark

$C.C.$ = Cupboard Clark's
$C.B.$ = Clark battery
P = Carbon resistance
τ = Box of 1·4 ohm coils
R = 10,000 ohms
L = Leclanche cell
T_1 = Principal thermometer (Baudin)
$T_2 T_3$ = Auxiliary thermometer

9 cm. in diameter and 4 cm. deep, whose weight was approximately 64 grammes. An experiment lasted ten minutes, during which about 0·56 grammes of silver were deposited, and the temperature of the water was raised about 2·2° C. All experiments were performed in the neighbourhood of 19·6°. The final result is the mean of six experiments which agree closely with each other.

The result of their investigation gives—

Capacity for heat of water at 19·1° C. on nitrogen scale = 4·1905 × 10⁷ ergs.

These experiments were conducted with the skill and accuracy which we necessarily associate with the name of Prof. Schuster; "Nevertheless," (I here again quote Prof. Ames) "there are several criticisms which may be offered to this research. There was only one voltameter used throughout, and none of the conditions were varied. The radiation corrections were most carefully considered, but no details are given of the stirring or of any correction for it. Griffiths in his investigation insists strongly on the need of thorough, not to say violent, stirring.

"These facts make the final result uncertain to an extent which it is difficult to estimate, but which probably is not large. If, as seems probable from the work of Kahle and Patterson and Guthe, the electro-equivalent of silver is 0·001119 instead of 0·001118, Schuster and Gannon's value for the specific heat at 19·1° becomes 4·189 × 10⁷; and, if a consequent error of one part in a thousand is made in the assumed value of the E.M.F. of their Clark cell, the corrected result is 4·185 × 10⁷."

It is also necessary to remember that Schuster and

Gannon did not trace the value of the secondary unit over any appreciable range of temperature, all their observations being confined to a rise of about 2·2° C. in the neighbourhood of 19° C.

I now wish to call your attention to the most recent enquiry of all, viz. that of Prof. CALLENDAR and Dr BARNES.

The main object of the investigation was not to determine the capacity for heat of water at any particular temperature; but to trace the variations in the capacity for heat with the temperature. A brief summary of the work was published in 1899, and in 1900 Dr Barnes presented to the Royal Society a full Report (not yet published)[1], together with an account of the revision of the work conducted by him, in which certain corrections were made for the eddies in the water and the effects of contained air.

I extract the following description of the apparatus from the summary by Prof. Callendar[2]:

" The general principle of the method, and the construction of the apparatus, will be readily understood by reference to the diagram of the Steady-flow Electric Calorimeter given in Fig. 14. A steady current of water flowing through a fine tube, is heated by a steady electric current through a central conductor of platinum. The steady difference of temperature between the inflowing and outflowing water is observed by means of a differential pair of platinum thermometers at either end. The bulbs

[1] An abstract will be found in *Proc. Roy. Soc.* 1900.
[2] B.A. Report, Dover, 1899.

of these thermometers are surrounded by thick copper tubes, which by their conductivity serve at once to equalise the temperature, and to prevent the generation of heat by the current in the immediate neighbourhood

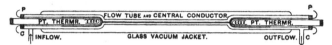

Fig. 14.   Diagram of Steady-flow Electric Calorimeter.

of the bulbs of the thermometers. The leads CC serve for the introduction of the current, and the leads PP, which are carefully insulated, for the measurement of the difference of potential on the central conductor. The flow tube is constructed of glass, and is sealed at either end, at some distance beyond the bulbs of the thermometers, into a glass vacuum jacket, the function of which is to diminish as much as possible the external loss of heat. The whole is enclosed in an external copper jacket (not shown in the figure), containing water in rapid circulation at a constant temperature maintained by means of a very delicate electric regulator.

" Neglecting small corrections, the general equation of the method may be stated in the following form :—

$$ECt = JMd\theta + H.$$

" The difference of potential E on the central conductor is measured in terms of the Clark cell by means of a very accurately calibrated potentiometer, which serves also to measure the current C by the observation of the difference of potential on a standard resistance R included in the circuit.

"The Clark cells chiefly employed in this work were of the hermetically sealed type described by the authors in the *Proc. Roy. Soc.* October 1897. They were kept immersed in a regulated water-bath at 15° C., and have maintained their relative differences constant to one or two parts in 100,000 for the last two years.

"The standard resistance R consists of four bare platinum silver wires in parallel wound on mica frames and immersed in oil at a constant temperature. The coils were annealed at a red heat after winding on the mica, and are not appreciably heated by the passage of the currents employed in the work.

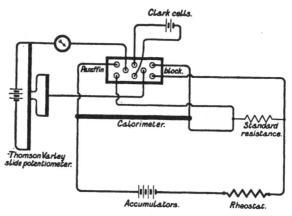

Fig. 15.   Diagram of the electrical connections.

"The time of flow t of the mass of water, M, was generally about fifteen to twenty minutes, and was recorded automatically on an electric chronograph reading to ·01 seconds, on which the seconds were marked by a standard clock.

"The letter J stands for the number of joules in one calorie at a temperature which is the mean of the range, $d\theta$, through which the water is heated.

"The mass of water, M, was generally a quantity of the order of 500 grammes. After passing through a cooler, it was collected and weighed in a tared flask in such a manner as to obviate all possible loss by evaporation.

"The range of temperature, $d\theta$, was generally from 8° to 10° in the series of experiments on the variation of J, but other ranges were tried for the purpose of testing the theory of the method and the application of small corrections. The thermometers were read to the ten-thousandth part of a degree, and the difference was probably in all cases accurate to ·001° C. This order of accuracy could not possibly have been attained with mercury thermometers under the conditions of the experiment.

"The external loss of heat, H, was very small and regular, owing to the perfection and constancy of the vacuum attainable in the sealed glass jacket. It was determined and eliminated by adjusting the electric current so as to secure the same rise of temperature, $d\theta$, for widely different values of the water-flow.

"The great advantage of the steady-flow method as compared with the more common method in which a constant mass of water at a uniform temperature is heated in a calorimeter, the temperature of which is changing continuously, is that in the steady-flow method there is practically no change of temperature in any part of the apparatus during the experiment. There is no correction required for the thermal capacity of the calorimeter; the external heat loss is more regular and certain, and there is no question of lag of the thermometers. Another incidental advantage of great importance is that the steadiness of the conditions permits the attainment of the highest degree of accuracy in the instrumental readings.

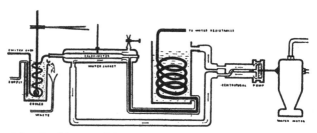

Fig. 16. Diagram of the arrangements for controlling the rate of flow of water.

" In work of this nature it is recognised as being of the utmost importance to be able to detect and eliminate constant errors by varying the conditions of the experiment through as wide a range as possible. In addition to varying the electric current, the water-flow, and the range of temperature, it was possible, with comparatively little trouble, to alter the form and resistance of the central conductor, and to change the glass calorimeter for one with a different degree of vacuum, or a different bore for the flow tube. In all six different calorimeters were employed, and the agreement of the results on reduction afforded a very satisfactory test of the accuracy of the method."

I am conscious of one difficulty regarding the expression of the results obtained by Callendar and Barnes, viz. the reduction of their numbers to the hydrogen scale, for, until the publication of the full paper, I am unable to give details regarding the standardisation of their platinum thermometers. As the method adopted was that published by Callendar and Griffiths in 1891[1], the thermometric scale is that of the constant pressure air-thermometer. Now, it is probable that over the range 0° to 100° C., this scale closely corresponds with that of the nitrogen thermometer, a conclusion verified to a certain extent by experiment.

In Table IV. (*infra*) the numbers under Col. 1 are the values of *J*, as given by Dr Barnes in the abstract[2] in which he gives an account of the further experiments conducted by him subsequently to the joint work by Professor Callendar and himself.

Under Col. 2 will be found the numbers I obtain by the conversion of Dr Barnes' values from the nitrogen to the hydrogen scale. For the purposes of this reduction I have used the Tables given by M. Chappuis[3], our leading authority on this matter.

---

[1] *Phil. Trans. Roy. Soc.* A. 1891.

[2] *Proc. Roy. Soc.* Nov. 1900.

[3] Mémoires du Bureau International. *Études sur le Thermomètre à Gaz*, p. 119.

TABLE IV.

Temperature	Col. I (Air scale)	Col. II (H scale)
5	$4\cdot2105 \times 10^7$	$4\cdot2130 \times 10^7$
10	4·1979	4·1999
15	4·1895	4·1912
20	4·1838	4·1851
25	4·1801	4·1805
30	4·1780	4·1780
35	4·1773	4·1774
40	4·1773	4·1769
45	4·1782	4·1776
50	4·1798	4·1785
55	4·1819	4·1806
60	4·1845	4·1828
65	4·1870	4·1854
70	4·1898	4·1881
75	4·1925	4·1912
80	4·1954	4·1946
85	4·1982	4·1979
90	4·2010	4·2014
95	4·2036	4·2050
Mean	4·1888	4·1887

On a later occasion I shall have to make further use of this Table. At present, for the purposes of comparison, we only require Callendar and Barnes' value on the nitrogen scale at 20°, i.e. $4\cdot184 \times 10^7$ ergs.

A summary of the results obtained by those observers, whose experiments we have examined, is given in the following Table.

TABLE V.

Capacity for Heat of Water per 1° of the N thermometer.

Name	Method	Standards	Results	Temperature
Joule ...	Mechanical		$4\cdot173 \times 10^7$	$16\cdot5°$
Rowland	,,		$\begin{cases} 4\cdot194 \\ 4\cdot186 \\ 4\cdot180 \\ 4\cdot176 \end{cases}$	$\begin{matrix} 10° \\ 15 \\ 20 \\ 25 \end{matrix}$
Reynolds and Moorby	,,		$4\cdot1833$	mean calorie
Griffiths	Electrical $\dfrac{E^2}{R} \cdot t$	Clark cell $=1\cdot4342$ International ohm	$\begin{cases} 4\cdot198 \\ 4\cdot192 \\ 4\cdot187 \end{cases}$	$\begin{matrix} 15° \\ 20 \\ 25 \end{matrix}$
Schuster and Gannon	E. C. t	Clark cell $=1\cdot4340$ El. Ch. E of Ag.$=0\cdot001118$	$4\cdot1905$	$19\cdot1°$
Callendar and Barnes	E. C. t	Clark cell $=1\cdot4342$ Ag.$=0\cdot001118$	$\begin{cases} 4\cdot198 \\ 4\cdot190 \\ 4\cdot184 \\ 4\cdot181 \end{cases}$	$\begin{matrix} 10° \\ 15 \\ 20 \\ 25 \end{matrix}$

The discrepancy between the values given in the above Table is, in reality, much less than would appear from a casual inspection. Before any real comparisons can be made, we must come to some conclusion regarding the variation in the capacity for heat of water, and when arriving at a decision on this matter we must also consider some evidence we possess which is independent of any determinations based upon the transformation of energy.

To-day I have devoted the greater portion of the time at our disposal to the consideration of experimental details and numerical values. In justification I will confess to holding the opinion that some teachers, in their anxiety to impart results, pay too little attention to the *methods* by which those results are obtained. A healthy scepticism, rather than a habit of comfortable belief, should, it appears to me, be cultivated by the seeker after natural knowledge. Text-books are not inspired, and teachers, above all, should learn to weigh the evidence and arrive at independent conclusions. We cannot rightly appreciate the authority of any natural law unless we have studied the experimental evidence upon which that law is based.

LECTURE IV.

In my last Lecture, I placed before you the numerical
values of the *mechanical equivalent* resulting from those
determinations which seemed to me the most trustworthy.
It is not possible, however, to compare these results and
arrive at a final decision unless we are able to trace those
changes in the value of the secondary thermal unit which
are due to the apparently capricious behaviour of water.

I wish here to make a distinction between two phrases
which are often employed (wrongly in my opinion) as
though they had the same meaning, viz., the capacity for
heat of water and the specific heat of water. I propose
to define the meaning of these terms as follows:—

By "capacity for heat of unit mass of water at $t°$"

or, more shortly, "the capacity for heat of water at $t°$," I indicate the number of ergs required to raise 1 gramme of water through 1° of the hydrogen scale at the temperature t on that scale.

By the specific heat of water, I denote "the ratio of the quantity of heat required to raise any mass of water through 1° at $t°$, to the quantity required to raise an equal mass of water through 1° at a standard temperature $θ°$, all temperatures being measured on the hydrogen scale."

Thus the capacity for heat, i.e. the value of C_t, is dependent on energy measurements, whereas the specific heat $σ_t$ can be obtained by comparison of quantities of heat only. True, if we know the capacity for heat of water at different temperatures, we can deduce the specific heat without further measurements since $\dfrac{C_t}{C_θ} = σ_t$, and there is little doubt that this method is the best, although the most laborious, way of finding $σ_t$. Nevertheless determinations of $σ$, not involving C, are of importance.

I would call your attention to the fact that the ratio of $\dfrac{C_t}{C_θ}$ is not affected either by inaccuracies in the magnitude of our electrical units, or by experimental errors in the determination of the value of the standards used by the observer, whereas the absolute value of C_t is dependent upon the accuracy of *all* the quantities involved. Thus, although electrical methods may be of secondary importance when our object is the determination of the numerical values of C_t, they are, for many reasons, of primary importance in the attempt to trace the changes in the value of $\dfrac{C_t}{C_\theta}$ consequent on changes in temperature.

I have already referred to the conservative nature of text-books. Take almost any one you please, and you will find tables of the specific heat of various bodies written to four or five places of decimals. Not content with this, the changes in its value are generally given by an equation of the form $\alpha + \beta t + \gamma t^2$, and occasionally even the third power is used. Now, I confess that I am sceptical as to the third significant figure of these specific values at a given temperature, and I refuse to give any credit at all to the fourth. Let us, however, assume that, at some temperature, the values given in these tables are correct, say to three significant figures; what importance can we attach to the values at other temperatures obtained from these equations? If you examine into the matter you will find that they are, in nearly all cases, dependent upon the assumed changes in the specific heat of water resulting from the extrapolation of Regnault's formula, or on Bosscha's reduction of Regnault's experiments[1]. At first sight this seems good enough, for it would be difficult to quote any higher authority where questions of accuracy are concerned. Regnault's experimental skill is admitted by all physicists and, therefore, the text-book writer goes ahead in all confidence, and peace and con-

---

[1] Usually given as $1 + \cdot00004\,t + \cdot0000009\,t^2$ (Regnault, *Mémoires de l'Acad.* xxi. p. 729, 1847) or $1 + \cdot00022\,t$ (Bosscha, *Pogg. Ann. Jah.* p. 549, 1876).

tentment prevail in the land. Let us enquire a little more curiously, however, and go back to Regnault's original papers. We find that with two exceptions Regnault performed no experiments concerning the specific heat of water below 107° C., and as these two were only made with a view of testing the working of the apparatus, he himself attached no importance to them. This really appears almost incredible; at all events it did so to me when I first enquired into the matter. The real facts are, I believe, as follows:—Regnault performed a series of determinations of the changes in the specific heat of water over the range 107° to 190° C. After discussing the results, he states what the nature of the variation between 0° and 100° would be if deduced by extrapolation of the experimental curve obtained at the higher range. Bosscha discussed Regnault's experiments, made several small corrections, found an equation which, in his opinion, more closely represented the experimental results over the range 107° to 190°, and then assumed that a similar expression held good down to 0°. Both Regnault's and Bosscha's equations are quoted by the text-books, and when once that has happened there is no escape.

Observe the consequences. Investigators who have performed experiments with the object of finding the specific heat of bodies at a certain temperature, and also the variation with temperature, have as a rule reduced the results to a standard temperature by these extrapolated values. Hence nearly all their conclusions require the revision which is rendered necessary by our knowledge that the changes in the specific heat of water between 0° and 100° differ both in magnitude and in direction from the changes which take place at higher temperatures.

No one who has not actually tried it can form any conception of the labour involved in attempting such revisions. In the majority of cases the process is at the best unsatisfactory, as the required data are not usually given; for they are not of the nature which, owing to his confidence in Regnault's supposed values, the writer of the paper has considered necessary. It is painful to contemplate the amount of good experimental work which has in consequence to be laid aside as useless.

Until 1879 Regnault's assumed values were universally accepted as Rowland was the first to supply sufficient *data* to justify the conclusion that, so far from increasing, the capacity for heat of water decreased with rise of temperature, at all events up to 30°. This result has, as we saw in my last Lecture, been confirmed by the work of Griffiths (range 13° to 27°), and Callendar and Barnes (range 1° to 99°).

The changes in the *specific heat* have been determined by Bartoli and Stracciati, and Ludin. In both cases the conclusions were arrived at by the method of mixtures, and are, therefore, independent of all energy measurements. Bartoli and Stracciati devoted nearly nine years to their investigation; they not only mixed water with water, but also with mercury and several metals. Their thermometry was based on the standards supplied by the Bureau International and their results are given in a formula containing the third power of *t*. They find a minimum

about 20°. The difficulties of this method are considerable, and a careful examination of their experimental
numbers leads to the conclusion that the discrepancies
between individual experiments are too great to allow
of our attaching much authority to their final values.

A very fine series of determinations by the same
method was made by Ludin in 1890. This investigator
was able to take advantage of many of the recent advances
in thermometric measurements, and his work is of a very
high order. Pernet has written a very full criticism of
this work and Ludin in his reply introduces a few corrections.

I now give a table which summarises the results of
the observers I have named. I have assumed 15° as the
standard temperature and in the case of the energy determinations (Rowland revised, Griffiths, and Callendar and
Barnes[1]) I have given the value of $\dfrac{C_t}{C_\theta}$ where θ is 15°. In
the cases of Bartoli and Stracciati and Ludin I have
expressed σ_t in terms of σ_{15} throughout.

This table is a very important one. I would first of
all call your attention to the comparatively close agreement between the values obtained by Rowland, Griffiths,
and Callendar and Barnes.

Now, remember how entirely different the methods
of experiment were in these three cases. In Rowland's,

---

[1] The numbers given in the last column of this table are deduced
from Col. II. Table IV. as it was necessary to express Barnes' results
in terms of the hydrogen scale before making a comparison with the
values obtained by the other observers.

TABLE VI.

Values of the Specific Heat of Water referred to that at 15° C. as unity.

Temperature by the hydrogen thermometer	Rowland (Day)	Bartoli and Stracciati (Pernet)	Griffiths	Ludin	Callendar and Barnes
0°	—	(1·0080)	—	(1·0051)	(1·0084)
1	—	—	—	—	—
2	—	—	—	—	—
3	—	1·0059	—	1·0035	—
4	—	52	—	31	—
5	(1·0042)	46	—	27	1·0052
6	1·0036	40	—	23	45
7	31	34	—	19	38
8	26	28	—	16	32
9	23	23	—	13	26
10	19	18	—	10	21
11	14	13	—	8	16
12	10	09	—	6	11
13	07	05	1·0006	4	07
14	03	02	03	2	03
15	1·0000	1·0000	1·0000	1·0000	1·0000
16	0·9996	0·9998	0·9997	0·9998	0·9997
17	93	97	94	97	94
18	90	96	91	96	90
19	86	95	88	95	88
20	83	94	85	94	85
21	81	93	82	93	83
22	79	93	79	93	80
23	76	94	76	92	77
24	74	95	73	92	75
25	72	97	70	93	74
26	71	98	0·9968	93	73
27	69	1·0000	—	94	72
28	69	02	—	94	71
29	68	05	—	95	70
30	67	10	—	96	69
31	67	1·0011	—	97	69
32	67	—	—	98	68
33	67	—	—	99	68
34	67	—	—	1·0001	67
35	0·9969	—	—	1·0003	0·9967

Values in brackets obtained by extrapolation.

the mechanical work was done against friction of water, the results as revised being dependent on the thermometry of the Bureau International. In Griffiths', the work done was measured by an electric current, the data being E and R, the thermometers mercury-in-glass instruments standardised by platinum thermometers and also by the standards of the Bureau International. In Callendar and Barnes', a very different method of estimating the heat developed by an electric current was adopted, viz. the use of a steady flow of water, the data being E and C, and the thermometry the differential platinum method.

We have seen that the values of C_θ are somewhat different in the three cases; but, on the other hand, this table shows that the values of $\dfrac{C_t}{C_\theta}$ are wonderfully concordant.

Now, observe the results of Bartoli and Stracciati and Ludin. The methods they adopted were similar in principle, although differing in detail; hence, one would have expected, *à priori*, that their results would have been in closer agreement than those obtained by the other observers. You see, however, that this is not the case; the differences between them are very marked (almost 3 in 1000 near 0°, and 1·4 in 1000 at 30°).

As it is unlikely that the skill and patience shown by Messrs Bartoli and Stracciati, and also by Ludin, can be exceeded, it is probable that the method of mixtures presents peculiar difficulties and uncertainties, which are absent in the energy determinations. Therefore, I think you will agree that if we take the mean value resulting from the observations of Rowland, Bartoli and Stracciati, and Griffiths over the range 13° to 26°, we shall be very near the truth. It is noticeable, by the way, that throughout the whole of this range the resulting values approximate most closely to Rowland's corrected results.

Before leaving this table it is interesting to notice how it illustrates the cosmopolitan character of scientific investigation and the importance attached to this matter by physicists of all nationalities.

The first column comes from Baltimore, the second from Pisa, the third from Cambridge, England, the fourth from Zürich, and the last from Montreal.

We are now, I think, in a position to reduce to some standard temperature the absolute values of C_t which were given in Table V.

Let us, by means of the curve representing our conclusions as to the changes in the specific heat over the range covered by these experiments, reduce them all to 20° C. on the N. scale. We get,

$$
\begin{array}{ll}
\text{Joule} \dotfill & 4{\cdot}169 \times 10^7. \\
\text{Rowland} \dotfill & 4{\cdot}180 \quad \text{,,} \\
\text{Griffiths} \dotfill & 4{\cdot}192 \quad \text{,,} \\
\text{Schuster and Gannon} \dotfill & 4{\cdot}189 \quad \text{,,} \\
\text{Callendar and Barnes} \dotfill & 4{\cdot}184 \quad \text{,,}
\end{array}
$$

I am afraid that in our search for the most probable value we must omit the number due to Joule. The doubt as to the exact conditions under which his thermometers were used renders the rejection advisable. (See p. 59, *supra*.)

As previously indicated, there is a considerable amount of evidence that the value assumed for the potential difference of a Clark cell is somewhat too high.

The value used in the electrical experiments was that found by Lord Rayleigh in 1884 as 1·4344 at 15° and redetermined by Messrs Glazebrook and Skinner in 1891, when they obtained 1·4342 at 15°. My own cells were compared with Lord Rayleigh's in a prolonged series of observations in 1892, and six of those cells were taken to Owens College and compared with Professor Schuster's in 1894; it would appear, therefore, that the *comparative* values of those cells are known with sufficient accuracy. The absolute values were obtained on the assumption that the electro-chemical-equivalent of silver is 0·01118.

On this point, I will again quote from the Report by Prof. Ames:

"In regard to the electrical standards, it must be observed that no meaning can be attached to the 'electro-chemical-equivalent' of silver, unless the construction and use of the voltameter are most carefully specified, and even then there is considerable doubt unless several instruments are used in series. This fact is well shown in the recent work of Richards, Collins and Heimrod at Harvard University, and of Merrill at Johns Hopkins University. The former deduce from a comparison of their porous-jar voltameter with other forms of instruments that the electro-chemical-equivalent with their instrument is 0·0011172 grams per sec. per ampère; while Patterson and Guthe with their instrument find 0·0011193. If, however, the same voltameter and the same method of use are adopted in the experiments on the electro-chemical-equivalent and in those on the E.M.F. of a Clark cell, the value of the latter is independent of the value assigned to the former. For this reason Kahle's value of the E.M.F. of the standard Clark cells of the Reichsanstalt (1·4325 volts at 15° C.) is probably correct. The Cavendish standard cell has been compared with the German ones; its resulting value is 1·4329 at 15° C. and the later investigations of Patterson and Guthe would reduce this to 1·4327. As Griffiths used the value 1·4342 and as the E.M.F. enters into the equation to the second power, the necessary correction would be almost exactly two parts in one thousand.

"For the method used by Schuster and Gannon, where both the E.M.F. and the current are measured, a correction may be accurately applied to the E.M.F., but not to the current, as it is not known what amount of silver one ampère should deposit in their voltameter; but if we assume that the correction in both cases is 1 in 1000, these results also would be reduced by 2 parts in 1000. The correction assigned is probably in the right direction.

"The cells used by Callendar and Barnes have not been compared with those of the Reichsanstalt, and no 'correction' can be applied with certainty[1]. The figures used above[2] are probably in excess."

[1] Here the value of the current was obtained by measurement of $\frac{E}{R}$, where R was a standard resistance, and it is somewhat difficult to estimate the probable effect of the change in the equivalent of silver in the absence of certain knowledge concerning the *comparative* value of Callendar and Barnes cells, in terms of the Rayleigh cell.

[2] i.e. a correction of 2 parts in 1000.

Assuming, therefore, the values of the E.M.F. of a Clark cell, and the electro-chemical-equivalent of silver deduced from the work of Kahle, Patterson and Guthe, we obtain the following table :—

Rowland[1]	$4 \cdot 180 \times 10^7$
Griffiths...........................	$4 \cdot 184$ „
Schuster and Gannon.........	$4 \cdot 181$ „
Callendar and Barnes.........	$4 \cdot 176$ „
Mean	$4 \cdot 1802$ at 20° N.

This close agreement between the mean and Rowland's number may be fortuitous, but, at the same time, the resulting coincidence between the values obtained by the mechanical and electrical methods renders it very probable that $1 \cdot 4328$ is more nearly the true potential difference of a Clark cell than the values $1 \cdot 4340$ to $1 \cdot 4342$ used in the original calculations[2].

Anyhow, *cæteris paribus*, we ought to attach far the greatest weight to the direct mechanical transformation, and, therefore, we can with confidence adopt the number $4 \cdot 180 \times 10^7$ as very near to the truth. This converted to the hydrogen scale at 20° $= 4 \cdot 181 \times 10^7$. If we assume as our standard change in temperature the rise from 17° to 18° (and I will presently explain the reason for this selection), we get (from Table VI.) the number $4 \cdot 184 \times 10^7$ ergs.

[1] $4 \cdot 181$ according to Waidner and Mallory (see Table V. *supra*) ; but as Day directly obtained the value $4 \cdot 181$ on the hydrogen scale, the probability is that $4 \cdot 180$ is the closer approximation on the nitrogen scale.

[2] For a further discussion of this matter see the Report of the Electrical Standards Committee, 1897.

In order to obtain the most probable values of C_t over the range 0° to 100° C. I propose to proceed as follows:

From Table VI. we can obtain the mean values of $\dfrac{C_t}{C_{15}}$ given by Cols. I. III. and V. over the range 0° to 35°. From 35° to 100° we must be guided by the observations of Callendar and Barnes (Table IV. Col. II.), and we can thus deduce the values of $\dfrac{C_t}{C_{17\cdot5}}$ over the whole range. The results of this reduction are given in Col. I. Table VII. (*infra*). Let us then assume the value $C_{17\cdot5} = 4\cdot184$ and deduce the value of C_t at other temperatures. The results are shown in Col. II. (*infra*).

Now, we have one test which we can apply to these numbers.

The result obtained by Reynolds and Moorby for " the mean thermal unit" was $4\cdot1833$ (*supra*, p. 75). A study of the tables given in their papers shows that the actual range was, on the average, from about 1°·2 or 1°·4 C. to 100°. An inspection of Table VII. will show that the rate of variation of $\dfrac{C_t}{C_\theta}$ is very rapid near 0° C., and that the probable value of $\dfrac{C_1}{C_{17\cdot5}}$ is about $1\cdot008$.

It is possible to make an approximate correction which, however, only raises Reynolds and Moorby's value to $4\cdot1836$. This result differs from the mean of the numbers in Col. II. Table VII. by less than 1 part in 2000. The correspondence is remarkable, and greatly increases the probability of the conclusions at which we have arrived.

TABLE VII.

Temperature on H. scale	Col. I. σ_t	Col. II. C_t
0°	(1·0083)	(4·219) × 10⁷
5	54	·206
10	27	·195
15	7	·187
20	·9992	·181
25	78	·176
30	75	·174
35	74	·173
40	73	·173
45	74	·173
50	77	·174
55	81	·176
60	87	·178
65	93	·181
70	1·0000	·184
75	7	·187
80	15	·190
85	23	·193
90	31	·197
95	40	·201
100	(1·0051)	(4·205)
Mean	1·00033	4·1854

It is possible that the methods of reduction may appear somewhat artificial. Remember, however, that even if we apply no correction for the possible errors in the electrical standards, we should probably have come to the same conclusion regarding the value of $C_{17\cdot5}$; for we should, in every case, have attached greater importance to Rowland's direct determinations than to any results obtained by indirect methods. Again, if when tracing the changes in C_t we had used Dr Barnes' original figures (without reduction to the hydrogen scale), the discrepancy between the values of the "mean unit" resulting from the work of Reynolds and Moorby, and Callendar and Barnes would not have been increased to as much as 1 in 1000. There is also a certain amount of indirect evidence, with which I will not trouble you, although it lends some support to our conclusions[1].

You will now understand the motives which dictated the proposal to regard the value of $C_{17.5}$ as the standard secondary unit, for it is almost coincident with the most probable value of the "mean thermal unit" over the range 0° to 100°.

This coincidence is a convenience, but not a necessity. Had the temperature 17° to 18° been in other respects unsuitable, I do not think it would have been wise to attach too much importance to this point. As it is, we are enabled to express the value of thermal measurements obtained by Bunsen's calorimeter in terms of the standard unit, and the accuracy is sufficient, for Prof. Nicholls has recently shown that the density of ice depends upon its rate of formation; the variations sometimes amounting to 2 parts in 1000[2]. Hence the experimental errors of such determinations are probably greater than those arising from the assumption that the secondary thermal unit at 17·5° C. is the same as the mean unit over the range 0° to 100°.

[1] See Griffiths, *Phil. Trans.* A. 1895, p. 321; Joly, *ibid.* p. 323; Griffiths, *Phil. Mag.* Nov. 1895.

[2] *Physical Review*, 1899.

The following is a summary of the conclusions to which our enquiry has led us:

1.   The *standard* (*secondary*) *thermal unit* is the energy required to raise 1 gramme of water from 17° to 18° C. on the Paris hydrogen scale, or one-fifth the amount required to raise it from 15° to 20° C. on the same scale.

2.   The value of this *standard secondary unit*

$$= 4 \cdot 184 \times 10^7 \text{ ergs.}$$
$$= 426 \cdot 5 \text{ kil. met. at Paris.}$$
$$= 1400 \cdot 0 \text{ ft.-lbs. at Greenwich.}$$

This unit expressed in terms of the F. scale at 63·5° F.

$$= 777 \cdot 7 \text{ ft.-lbs. at Greenwich.}$$

3.   The value of the *secondary unit* at other temperatures over the range 0° to 100° approximates closely to that given in Col. II. Table VII.

4.   The *mean thermal unit* over the range 0° to 100° C. may be taken as identical with the *standard unit* as defined in (1) *supra*.

The above conclusions are practically the same as those presented by me in a Report to the Paris Congress in 1900; although, as Dr Barnes had not then published the results of his revision of his own and Callendar's work, some small modifications have now been made, but they nowhere amount to more than 1 in 2000.

If, as I hope, thermal quantities are in future expressed in terms of the values given in Table VII., we shall at all events get rid of many of the difficulties which have hitherto hampered and perplexed all students of thermal measurements[1].

---

[1] See Appendix III.

The first law of thermodynamics is usually enunciated as follows. "When work is expended in producing heat the quantity of heat generated is proportional to the work done," and the evidence at our disposal is not only sufficient to establish the existence of such a proportion but also enables us to assign to it a definite numerical value.

The enunciation of this law was an event not only important in itself but far-reaching in its consequences. You remember that—from the time of Newton to that of Joule—the one thing wanting to justify a belief in the indestructibility of energy was experimental proof of a definite relation between work done in producing heat and the heat developed. The discovery of this missing link completed the chain of evidence, and the truth of the principle of the Conservation of Energy at once became apparent. The effects of a general recognition of this truth were soon evident in every branch of science, for it was found that phenomena hitherto regarded as isolated or disconnected were in many cases related as closely as cause and effect. To-day the validity of that principle is acknowledged alike by the physicist, the chemist, the biologist and the engineer, and one and all reject such explanations of natural phenomena as can be shown to conflict with this, the greatest and most fruitful generalization of modern physics.

It is possible that when speaking of "different forms of energy" we are but emphasizing unnecessary distinctions. If so we may state the principle as follows: "the energy of motion is indestructible," in other words, all transmutations of energy are really but the presentation of motion under differing circumstances. The existence of potential energy may appear a difficulty. Suppose a body shot upwards and caught at its highest point; what has become of its energy of motion? The supposition that we have transferred its initial motional energy to the ether naturally presents itself. If a body when accelerated receives energy of motion *from* the ether, and if when retarded it communicates energy of motion *to* the ether, then by potential energy we mean energy of motion of ether external to the body; by kinetic energy, motion of the body, or in the body, itself.

Perpetual motion (not, as Prof. Tait well remarks, "*the* perpetual motion") is a law of nature, and it is possible that the doctrine of the indestructibility of motion may be regarded as of equal validity to that of the indestructibility of matter; or again, the two statements may after all be but different aspects of one and the same proposition.

Although the second law involves no numerical determinations, we should remember that it, like the first, is a purely experimental one.

"That it is impossible to derive mechanical effect by means of heat obtained from any portion of matter by cooling it below the temperature of the coldest of the surrounding bodies" is a conclusion dependent simply upon experience, like all other natural laws. It should always be borne in mind, however, that this law when thus enunciated, is only applicable in those cases in which the working substance has been taken through a complete cycle, and has therefore been restored to its initial condition. For example (as previously pointed out, p. 17) we can derive mechanical effect by allowing a gas to expand against pressure, and, as a consequence, its temperature may fall below that of surrounding bodies. This operation, however, is not a complete cycle, for work has been done at the expense of the internal energy of the working substance. We cannot draw any legitimate conclusion, concerning the relation between the heat supplied to the system and the work obtained from it until the working substance has been restored to its initial state.

I have an impression that students rarely appreciate the importance, alike to the physicist and the chemist, of this second law. By means of it we can not only test the validity of our conclusions regarding operations involving transference of energy, but we can also predict phenomena and verify by experiment the truth of our predictions. One beautiful application is, of course, known to you all, viz. J. Thomson's calculation of the effect of increase in pressure upon the melting-point of ice. The usual proof is of a mathematical nature; but, if you will permit me, I should like to show that we can arrive at the same conclusion by going through a series of operations somewhat similar to those already performed by means of Carnot's engine.

Let the hot and cold bodies X and Y in Fig. 2, p. 35, be replaced by two equal vessels R and Q (Fig. 17) similar in form to V, their walls being made of an adiabatic substance except where they come in contact with the sides of V at the shutters L and K, which you remember

can be changed at will from adiabatic to perfectly conducting faces, and thus, when open, permit the transference
of heat from one vessel to another, although the system as
a whole is thermally isolated.

Let Q contain ice and water, then if the pressure on p_2
is that of one atmosphere, the temperature within Q will
be 0° C.

Fill R also with ice and water, and let a heavy weight
W be supported by its piston p_3, so that the pressure in
R always exceeds that in Q.

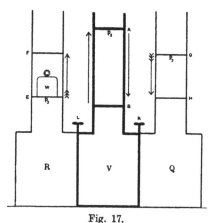

Fig. 17.

Open K so that the substance in V assumes the
temperature of Q (*i.e.* 0° C.), and let its piston (p_1) stand
at A. Now force p_1 downwards, say to B; the heat
produced by the work thus done will pass into Q, melting
some ice, but not altering its temperature. This will
cause the contents of Q to diminish in volume and the
piston p_2 will descend through some distance GH. Now

close K and open L and allow p_1 to rise until it is again at A. Heat will be taken from R during this operation and some ice must be formed in R. The piston p_3 must therefore rise, lifting the weight W.

Now, *if we assume that the temperature of R during the formation of ice remained at* $0°$ C., then the working substance in V must have remained at the same temperature throughout the cycle, and therefore, as its piston has returned to its original position at A, no external work has been done by its contents.

Again, since the pressure on p_3 is greater than that on p_2, the work done *on* Q could only equal that done *by* R if the piston p_2 had descended through a greater distance than that ascended by p_3. This is not, however, possible; for, if so, a greater volume of ice must have been melted in Q than was formed in R and thus the system would have received heat without disappearance of other forms of energy. Hence (on the above hypothesis that R remained at $0°$ C.) the system has done work by the transference of heat from one body to another at the same temperature. This, as the second law tells us, is contrary to all experimental evidence.

We are thus forced to the conclusion that during the cycle sufficient work must have been done *on* V to restore the balance of energy. Therefore, the work done by the substance in V during an upward, was less than that done on it during a downward, stroke. That is, the pressure, and therefore the temperature, of V was diminished by shutting K and opening L; hence the temperature of the mixture in R was lower than that in Q, *i.e.* below $0°$ C.

The same method of reasoning shows that if a body contracts on solidification, its freezing-point must be raised by pressure.

Many other similar applications suggest themselves. For example, the volume of a saturated solution of ammonium chloride in water exceeds that of the water and salt when separate. Hence the application of great pressure would cause the formation of crystals. On the other hand, the volume of a saturated solution of copper sulphate in water is less than the volume of its constituents. Hence the effect of greater pressure must be the disappearance of crystals. I presume, therefore (although I do not know of any experimental evidence), that if we took two long columns of these solutions we should perceive the crystals of ammonium chloride form at the bottom, while those of copper sulphate would form near the top.

In estimating the importance of this second law, we must remember that by means of it we were able to complete the demonstration of the truth of the statement that a reversible engine is the best possible engine, and that, therefore, all reversible engines are equally efficient, and it is upon this conclusion that our true conception of an absolute temperature scale depends.

Our scientific knowledge of to-day is a great building containing many chambers. The whole edifice has been erected upon certain fundamental piers of which not the least important is this conception of an absolute temperature scale, and if this pillar were removed it would be difficult to set any limit to the effect upon the stability of the whole structure.

Reflect for a moment on one of the consequences which results from our definition of an absolute scale, viz. that if a source is at an absolute temperature θ and our sink, or condenser, at temperature t, then the efficiency of a perfect engine working between these temperatures is $\dfrac{\theta - t}{\theta}$. That statement alone not only gives the engineer complete knowledge of the limitations under which he works, but also the ideal limit towards which he should ever struggle, although he can never attain to it.

"Never" may appear to be a dangerous word to use; but in this case it is allowable. So long as the conditions of the universe remain as we now find them, the word is not only allowable but necessary. Let us consider the matter more fully. You remember that when we traced the various steps of the Carnot cycle, we saw that the area representing the work done during that cycle was bounded by isothermal and adiabatic lines. Now, we have not as yet discovered either a perfect conductor or an adiabatic substance; thus during a downward stroke the contents must always be at a higher temperature than the walls and, therefore, the pressure must exceed that shown by the isothermal until we have come to rest at the end of the stroke. Conversely, during an upward movement the contents must be cooler and hence the pressure less. During a forward cycle, therefore, the area of the work-space is diminished and during a reverse cycle it is increased[1].

[1] The dotted lines in Fig. 3 (p. 38, *supra*) indicate the effect upon the area of the work-space during a forward cycle.

The only way of meeting this difficulty would be for the movement of our piston during the isothermal strokes to be infinitely slow. Again, during the so-called adiabatic strokes some heat must enter or leave the cylinder in any given interval of time, as we do not possess any adiabatic substance. These strokes would, therefore, have to be made with infinite rapidity. Hence, when the valves communicating with either the source or the condenser are open, our engine must move with infinite slowness; and when those valves are closed, with infinite rapidity. Also all movements must be frictionless if it is to play the part of a reversible engine. Here, I think, are enough impossibilities to justify the use of the word " never."

We know that under the best possible conditions, the elevation of heat to other forms of energy can never be more than partial; whereas the complete degradation of the mechanical to the heat form of motion is constantly taking place. Again, such partial restoration as we have is only feasible where differences of temperature exist and where, by its very occurrence, it tends to destroy those differences of temperature, and thus the *available* or *free energy* of a system always tends to become a minimum. The conclusion is inevitable. All energy must ultimately be degraded to the heat form and become so distributed that all matter is at one temperature.

It is not possible at the close of such a course of lectures as this to give proper consideration to the grand generalisation known as the Dissipation of Energy—a generalisation perhaps as important as that of con-servation.

Even as the Conservation of Energy recalls the name
of Joule, so does the Dissipation of Energy suggest the
name of William Thomson, and I cannot more fitly
summarise this portion of our subject than by the following quotation from the writings of Lord Kelvin :

"Exhaustive consideration of all that is known of the
natural history of the properties of matter, and of all
conceivable methods for obtaining mechanical work from
natural sources of energy, whether by heat-engines, or by
electric engines, or water-wheels, or tide-mills or any other
conceivable kind of engine, proves to us that the most
perfectly designed engine can only be an approach to the
perfect engine; and that the irreversibility of actions
connected with its working is only part of a physical
law of irreversibility according to which there is a uni-
versal tendency in nature to the dissipation of mechanical
energy; and any *partial restoration* of mechanical energy
is impossible in inanimate material processes and is
probably never effected by means of organised matter,
either endowed with vegetable life, or subject to the
will of an animal.

"The doctrine of the 'Dissipation of Energy' forces
upon us the conclusion that within a finite period of
time past, the earth must have been, and within a finite
period of time to come must again be, unfit for the
habitation of man as at present constituted, unless operations have been, and are to be, performed, which are
impossible under the laws governing the known operations
going on at present in the material world[1]."

[1] *Lectures and Addresses*, Lord Kelvin, Vol. II. p. 469.

In this necessarily brief survey, I have endeavoured to set before you, as fully as time permitted, the extent of our present knowledge regarding one of the leading principles of natural science. I will now venture to recall to your memories the following quotation from my first Lecture:

"The recent history of the establishment and development of the Conservation of Energy is a steady record of progress; but that progress partakes rather of the nature of an accurate survey of country whose main outlines are already familiar, than of any advance into unknown territory.

"It is not on that account, however, of less consequence; for settlement is often as true an indication of progress as discovery itself."

This statement has, I hope, been justified by the facts to which I have called your attention during these lectures.

I have only been able to give you faint outlines, even of the works selected for consideration; the details must be filled in by direct study of the original papers. Enough has been said, however, to indicate the strength of the evidence that has been gradually collected by many observers in different lands.

I doubt if so much time, thought, and experimental skill have been devoted to the determination of any other physical quantity. Such labours, however, have not been in vain, for at the commencement of this new century we are in a position to speak with some confidence regarding the value of what I do not hesitate to call the most important of all natural constants.

APPENDIX I.

THE THERMAL UNIT.

In December, 1895, I received from Professor Rowland a letter regarding the question of the selection of a secondary unit. In answer to a request, he then gave me permission to publish that letter, in whole or in part, and the portion here given will be found in the Reports of the Electrical Standards Committee[1].

I now reproduce it, not only on account of the importance of the opinions therein expressed, but also because it possesses at the present time a sad interest, as a record of the great physicist who has so recently passed away.

> "JOHNS HOPKINS UNIVERSITY,
> *December* 15, 1895.

"As to the standard for heat measurement, it is to be considered from both a theoretical as well as a practical standpoint.

The ideal theoretical unit would be that quantity of heat necessary to melt one gramme of ice. This is independent of any system of thermometry, and presents to our minds the idea of quantity of heat independent of temperature.

Thus the system of thermometry would have no connection

---

[1] *B.A. Report*, Liverpool, 1896.

whatever with the heat unit, and the first law of thermo-dynamics would stand, as it should, entirely independent of the second.

The idea of a quantity of heat at a high temperature being very different from the same quantity at a low temperature, would then be easy and simple. Likewise we could treat thermo-dynamics without any reference to temperature until we came to the second law, which would then introduce temperature and the way of measuring it.

From a practical standpoint, however, the unit depending on the specific heat of water is at present certainly the most convenient. It has been the one mostly used, and its value is well known in terms of energy. Furthermore, the establishment of institutions where *it is said* thermometers can be compared with a standard, renders the unit very available in practice. In other words, this unit is a better practical one at present. I am very sorry this is so, because it is a very poor theoretical one indeed.

But as we can write our text-books as we please, I suppose that it is best to accept the most practical unit. This I conceive to be the heat required to raise a gramme of water 1° C., on the hydrogen thermometer at 20° C.

I take 20° because in ordinary thermometry the room is usually about this temperature, and no reduction will be necessary. However, 15° would not be inconvenient, or 10° to 20°.

As I write these words I have a feeling that I may be wrong. Why should we continue to teach in our text-books that heat has anything to do with temperature? It is decidedly wrong, and if I ever write a text-book I shall probably use the ice unit. But if I ever write a scientific paper of an experimental nature, I shall probably use the other unit."

APPENDIX II.

APPROXIMATE METHODS OF DETERMINING THE
MECHANICAL EQUIVALENT.

As stated at the close of the second Lecture, I give the following notes in the hope that they may be found useful by teachers.

Experiment I. A very thin-walled brass tube, diameter about 9 cm., length 110 cm., was surrounded by wrappings of asbestos and then firmly fixed within a strong iron cylinder[1]. The ends of the tube were closed by brass plates and those of the cylinder by iron plates, the end spaces being also filled with asbestos.

A quantity of lead shot was placed within the brass tube, the distance from the surface of this shot to the opposite end of the tube being 102 cm. when the tube was vertical.

A small thin-walled re-entering tube, whose diameter was just sufficient to contain the bulb of a thermometer, was so fixed that its closed end was in the middle of the shot, while its open end was in the flat surface at the base of the iron cylinder. The temperature of the shot was determined by inserting the thermometer in this tube at the beginning and end of an experiment.

The iron cylinder was clasped at its centre by a band fixed to the end of a horizontal axis, and at the further end of that axis was placed a cross-bar terminating in handles. The bearings of the axis were supported on a firm stand. If the

[1] This apparatus is merely a more elaborate (and therefore probably inferior) form of that described in Edser's *Heat*, p. 272.

cylinder was rotated with sufficient rapidity the shot did not fall, even when the loaded end was uppermost; but if the motion was then arrested, the shot fell through 102 cm. on to the flat plate at the lower end of the brass tube.

A projecting bar, controlled by springs, was fitted on to the stand in such a manner as to arrest the motion of the cylinder whenever it assumed a vertical position. The energy due to the motion of the cylinder was chiefly expended in warming the surfaces rubbed by these springs.

If h is the distance through which the shot falls in each stroke, then its kinetic energy when reaching the lower end is $\frac{1}{2}mv^2$, i.e. mgh. Thus the work done in n strokes is $n \cdot mgh$. Let the rise in temperature of the shot be θ and σ the specific heat of the lead. Then heat developed $= \sigma m\theta$.

$$\therefore \ \frac{n \cdot mgh}{\sigma m\theta} = J = \frac{ngh}{\theta\sigma}.$$

In the actual experiment performed at the close of the lecture, 50 strokes were given and $h = 102$ cm.

$$\therefore \ ngh = 5,000,000 \text{ approximately.}$$

The observed rise in temperature was $3\cdot4°$ C. and we may take $\sigma = \cdot033$.

Therefore

$$J = \frac{5 \times 10^6}{3\cdot4 \times \cdot033} = 4\cdot46 \times 10^7.$$

In this case it would be difficult to estimate the loss of heat by radiation, etc., or to allow for the thermal capacity of the tube. Even if such corrections were possible they would have little significance, and their insertion would impart a false appearance of accuracy to the results.

The experiment is useful educationally if regarded rather as an *illustration* of the conversion of kinetic energy into heat, rather than as an attempt to obtain a numerical value of the equivalent.

Experiment II. In all essential points the following method is similar to that adopted by Joule (in his later experiments) and by Rowland.

Although results obtained by this apparatus cannot of course (in the absence of elaborate precautions) lay claim to a high order of accuracy, yet, in the hands of a capable student, the probable error should not exceed 1 to 2 per cent. A higher order of accuracy would be useless, for the errors in thermometry would, under ordinary circumstances, be of the same order. In the example given (*infra*) I have reason to believe that the error of the thermometer is probably of the order of 1 per cent. over the range 10° to 35° C.

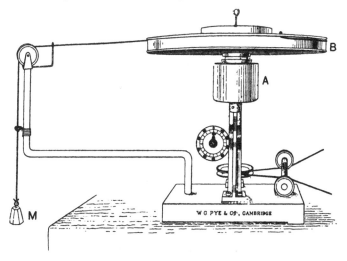

Fig. 18.

1. Description of the apparatus[1]. In the machine

[1] This account of the apparatus has, by the kind permission of Mr G. F. C. Searle, been copied from the note-book prepared by him for the

used for this experiment, a vertical spindle carries at its upper end a brass cup *A* (Fig. 18). Into an ebonite ring concentric with *A* there fits tightly one of a pair of hollow truncated cones. The second cone *C* (Fig. 19) fits into the first one and is provided at its upper edge with a pair of steel pins which correspond to the two holes in a grooved wooden disc *B* (Fig. 18). In the experiment, the disc *B* prevents the inner cone from revolving when the spindle and the outer cone revolve, and the friction between the two cones gives rise to heat. A cast-iron ring, resting on the disc and fixed by two pins, serves to give a suitable pressure between the cones.

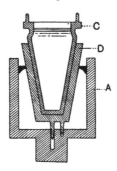

Fig. 19.

A brass wheel is fixed to the spindle (Fig. 18), and by a string passing round this wheel, and also round a handwheel, motion is imparted to the spindle. A pair of guide pulleys prevents the string from running off the wheel. Above the wheel is a screw cut upon the spindle which moves a cogwheel of 100 teeth, which makes one revolution for each 100 revolutions of the spindle.

To the base of the apparatus one end of a cranked steel

use of students at the Cavendish Laboratory. A few trivial alterations have been made therein in consequence of certain slight modifications introduced into the apparatus used by me at Leeds.

rod is attached and the rod can be fixed in any position by a nut beneath the base. The other end of the rod carries a cradle in which runs a small guide pulley, this pulley being on the same level as the disc. The cradle turns freely about a vertical axis.

A fine string (plaited silk fishing line) is fastened to the disc and passes along the groove in its edge; it then passes over the pulley and is fastened to M, a mass of 200 or 300 grammes. On turning the handwheel it is easy to regulate the speed so that the friction between the cones just causes M to be supported at a nearly constant level.

Through unskilful driving of the handwheel, the string may slip off the guide pulley, and in consequence off the disc; the mass M may also be wound up over the guide pulley. To prevent the string from running off the guide pulley, a stiff wire with an eye is fixed to the cradle, in such a manner that the eye is on the same level as the groove of the pulley and about 5 cm. from the axle of the pulley towards the disc. If the string is passed through this eye it will always turn the cradle so that the string runs fairly over the pulley. In order to prevent the mass M from being wound up over the pulley an eye is fixed to the steel rod, and the string supporting M passes through this eye. With these arrangements it is impossible either to throw the string off the guide pulley or to wind M up over the pulley.

2. Setting up the apparatus. Two tables will generally be required. The frictional machine is firmly clamped to one table, and the handwheel is clamped to the other table at a distance of 10 or more feet. Care must be taken that the driving string runs properly, without any risk of slipping off the handwheel. The steel rod is also fixed in a convenient position.

A thermometer is hung from a support so that it passes through the central aperture in the disc, and almost touches the bottom of the inner cone. The thermometer should also pass through the hole in the stirrer.

The string supporting M should be of such a length, that when as much as possible has been unwound from the disc, M is not quite in contact with the floor.

Before putting the cones together the rubbing surfaces must be carefully cleaned, and then four or five drops of oil must be put between them; the bearings of the spindle and guide pulleys should also be oiled.

3. Method of experimenting. The cones, cleaned and oiled, are weighed, together with the stirrer. The inner cone is then filled with water up to about 1 cm. from its edge, and the system is again weighed. The water should be at the temperature of the room. The apparatus is now put into working order, the counting wheel having previously been brought to its zero position. One observer, X, takes his place at the handwheel, and a second observer, Y, at the machine. After the initial temperature of the water has been carefully observed, the operator X turns the handwheel at such a rate, that the mass M is raised so far from the floor that the string supporting M is a tangent to the edge of the disc. If the string is not a tangent to the disc the moment of the tension about the axis of revolution is seriously diminished. The observer Y stirs the water and notes the temperature at the end of each 100 revolutions of the spindle. He gives a signal as each 100 revolutions is completed, and X notes the time upon a watch. After Y has recorded the temperature upon a sheet of paper previously ruled for the purpose he also records the time observed by X. *Very* accurate readings of these temperatures and times are difficult to make and are not necessary. When M is 200 grammes the temperature will rise about 0·8° C. for each 100 revolutions of the spindle in the case of apparatus similar to that used in the lecture. It will be found that the time occupied by 100 revolutions of the spindle diminishes as the temperature rises; this effect is due to the diminution of the viscosity of the oil between the cones, consequent upon the rise of temperature.

After about 1000 revolutions have been made by the spindle the motion is stopped, and the highest temperature shown by the thermometer is carefully read. The index of the counting wheel is also observed, and from this reading and the number of complete revolutions made by the counting wheel the exact number of revolutions made by the spindle is ascertained.

Without disturbing the apparatus the water is allowed to cool, and observations of the temperature are taken at the end of each one or two minutes till the water is only slightly (2 degrees or so) above the temperature of the room.

4. Calculation of the correction for cooling. If there had been no loss of heat during the time that the apparatus was in action, the difference between the final and initial temperatures of the water could be used in the calculation, without any correction. The correction necessary to allow for the loss of heat in the actual case is ascertained in the following manner. From the observations taken, a curve is plotted, (the abscissae denoting time, the ordinates temperature) showing how the temperature increased with the time. On account of the increase of speed, due to the diminution in the viscosity of the oil, the curve is concave upwards.

From the observations on the cooling of the water a second curve is drawn with time as abscissa, and temperature as ordinate; and from it the rate of cooling in degrees per minute at any particular temperature is determined by the slope of the tangent to the curve. It is best not to actually *draw* the tangent. If a triangular "drawing square" *ABC* be adjusted so that one of its edges *AB* touches the curve at the desired point, and if one of the other sides, as *AB*, be made to slide along a straight edge, AB can be moved parallel to itself until it passes through an intersection of the lines ruled on the squared paper. It is now easy to read off along the edge of the square the number of degrees lost in 10 minutes. Dividing this number by 10, the rate of cooling, in degrees per minute, is obtained for the particular temperature at which the tangent was taken.

A third curve is now constructed. From the first curve the temperatures of the water at the end of 1, 2, 3...minutes are determined, and by the second curve the rates of cooling at these temperatures are determined. The third curve is drawn with the time as abscissa and the corresponding rate of cooling as ordinate. The origin is one point on the curve, since the water was initially at the temperature of the room. The area included between the line of times, the curve and the ordinate corresponding to the time when the highest temperature was noted, represents the total loss of temperature during the experiment. If 1 inch (or cm.) on the squared paper corresponds to p minutes, and 1 inch (or cm.) corresponds to q degrees per minute, then each square inch (or square cm.) represents pq degrees.

(If the original temperature was below that of the room experiments must be made to determine the rate of heating for the lower temperatures. In this case part of the curve will lie below the axis of time, and the area of that part is to be taken as *negative*.)

5. Calculation of the Mechanical Equivalent of Heat.
When the spindle has made n turns, the work which has been spent in overcoming the friction between the two cones is the same as if the outer cone had been fixed and the inner one had been made to revolve by the mass M grammes. In the latter case M would have fallen through $2\pi nr$ cm., where r centimetres is the radius of the groove of the wooden

disc. Hence the total work spent upon overcoming friction is $2\pi n r M g$ ergs.

Let W be the mass of the water in grammes.

    ,, w ,,   ,,   ,,   cones and stirrer.

The specific heat of the brass, if not previously determined, may be taken as ·095, and thus the system of cones, stirrer and water is thermally equivalent to $(W + ·095w)$ grammes of water.

(If the thermometer has a large bulb it is necessary to take account of its water equivalent. Its water equivalent is found by heating the thermometer and plunging it while hot into a small vessel containing a known quantity óf water at a known temperature. From the rise of temperature the water equivalent is calculated. Its value must be added to $W + ·095$ in the equation for J.)

Let θ be the observed rise of temperature, and ϕ the calculated total loss of temperature during the time for which the experiment was in progress. Then the total number of thermal units produced is $(W + ·095w)$ $(\theta + \phi)$ water gramme degrees.

If J denote the number of ergs of work which must be spent to produce one thermal unit, we have

$$J = \frac{2\pi n r M g}{(W + ·095w)\,(\theta + \phi)}.$$

As an example I give the details of the following experiment, which was performed before the lecture; as it would have been impossible to obtain the cooling curve in the time available during, or after, its close. The working of the apparatus was however exhibited, and the experimental numbers previously obtained were given; with a request that the results should be worked out before the following meeting.

DETAILS OF AN EXPERIMENT PERFORMED
MARCH 9TH, 1901.

Weight of the two brass cones and stirrer = 164·1 grms.
    ,,   ,, contained water   = 21·35 ,,
Suspended mass   = 300 ,,
Circumference of wheel   = 78·40 cm.
Specific heat of brass cones and stirrer   = ·095
Capacity for heat of thermometer   = 0·1

During experiment the external temp. rose from 10·2° to 10·6° C.

TABLE A.	TABLE B.
Rate of rise during rotation.	*Observations on rate of cooling after experiment.*

Time		Temperature C.	Time		Temperature C.
Minutes	Seconds		Minutes	Seconds	
0	0	10·4	18	13	33·0[1]
0	50	11·0	18	20	33·1
2	25	12·0	18	50	32·6
3	57	13·0	19	32	32·0
5	23	14·0	20	39	31·0
6	45	15·0	21	57	30·0
7	58	16·0	23	20	29·0
9	12	17·0	25	2	28·0
10	15	18·0	26	43	27·0
11	8	19·0	28	42	26·0
12	0	20·0	30	44	25·0
12	45	21·0	32	59	24·0
13	28	22·0	35	56	23·0
14	8	23·0	39	10	22·0
14	46	24·0	42	42	21·0
15	22	25·0	46	10	20·0
15	50	26·0	50	42	19·0
16	11	27·0	55	42	18·0
16	37	28·0	61	15	17·0
17	2	29·0			
17	23	30·0			
—	—.	31·0			
18	0	32·0			
18	13	33·0[1]			
18	20	33·1[2]			

[1] Rotation stopped.

[1] Rotation stopped.
[2] Highest temperature reached.

Total number of revolutions = 1854.

REDUCTION OF THE RESULTS.

Capacity for heat of cones, thermometer and water = 37·05.
Observed rise in temperature   = 22·7° C.
Loss of temperature by radiation, etc.   = 4·52.
Hence total heat evolved = 37·05 × 27·22 thermal units.
Work done = 1854 × 78·4 × 300 × 981 = 4277 × 10⁷ ergs.

$$\therefore\ J = \frac{4277 \times 10^7}{37\cdot05 \times 27\cdot22} = 4\cdot241 \times 10^7.$$

This is the mean value over the range 10 to 33° C. by a mercury-in-glass thermometer. The value would probably be raised by reduction to the H scale.

APPENDIX III.

COPY OF RESOLUTIONS PASSED BY THE ELECTRICAL
STANDARDS COMMITTEE OF THE BRITISH ASSOCIA-
TION IN 1896.

Propositions.

" I.   The fundamental thermodynamic unit of heat is
10^7 ergs, to which unit the name *joule* has already
been given by the Electrical Standards Committee
of the British Association.

For many practical purposes heat will continue to be
measured in terms of the heat required to raise a measured
mass of water through a definite range of temperature.   If
the mass of water be 1 gramme, and the range 1° C. in the
neighbourhood of 10° C., then the number of *joules* required
will be approximately 4·2.   It will be convenient to fix upon
this number of *joules* as a thermometric unit of heat, and to
state—

"ₗII.   The thermometric unit of heat is 4·2 *joules*.

According to the best of the existing determinations (see
Mr Griffiths' paper already quoted), this is the amount of heat
required to raise 1 gramme of water from 9°·5 C. of the scale
of the hydrogen thermometer to 10°·5 of that scale.

Accordingly for the present [or until the year 1905?] a third proposition would be—

"III. The amount of heat requisite to raise the temperature of 1 gramme of water 1° C. of the scale of the hydrogen thermometer from 9°·5 to 10°·5 C. of that thermometer, is equivalent to one thermometric unit of heat.

In case further research should show that this statement is not exact, the definition could be adjusted by a small alteration in the mean temperature at which the rise of 1° takes place. The definitions in I. and II. would remain unaltered."

As the conclusion given in Proposition III. of this Report differs considerably from that given on page 110, I take this opportunity of calling attention to the following points:

(1) The revision of Rowland's results, the determinations of Reynolds and Moorby, and of Callendar and Barnes had not been accomplished at the time the Committee presented the above Report.

(2) It will be seen (*supra*) that Proposition III. was regarded by the Committee as a provisional statement.

(3) The effect of the subsequent increase in our knowledge is to shift the probable temperature at which the *capacity for heat* of water is 4·2 *joules* to about 7·5° C. This temperature is certainly inconveniently low and therefore it is desirable that a change should now be made in Proposition III.

Printed in the United States
By Bookmasters